INVERSE COORDINATION CHEMISTRY

A NOVEL CHEMICAL CONCEPT

IONEL HAIDUC

EDWARD R.T. TIEKINK

Published by Sunway University Press
An imprint of Sunway University Sdn Bhd

No. 5 Jalan Universiti
Bandar Sunway
47500 Selangor Darul Ehsan
Malaysia

press.sunway.edu.my

ISBN 978-967-5492-18-1

Perpustakaan Negara Malaysia   Cataloguing-in-Publication Data

Haiduc, Ionel
INVERSE COORDINATION CHEMISTRY : A NOVEL CHEMICAL CONCEPT /
Ionel Haiduc, Edward R.T. Tiekink.
ISBN 978-967-5492-18-1
1. Chemistry, Inorganic.
2. Chemistry.
I. Tiekink, Edward R. T.
II. Title.
546

Edited by Sarah Loh, Oileen Chin
Designed by Rachel Goh
Typeset by Pauline Loo
Printed by Valley Printers, Malaysia

The cover illustration represents the molecular structure of the inverse coordination complex, $(\mu_6\text{-}C_6S_6)[AuPPh_3]_6$, described in H.K. Yip, A. Schier, J. Riede and H. Schmidbaur, *J. Chem. Soc., Dalton Trans.*, 1994, 2333–2334.

Contents

Preface

Inorganic chemists have traditionally described coordination complexes in a metal-centric manner, whereby the metal atom is the focus and is usually embedded in an organic environment or surrounded by inorganic species—like halogens, water, or ammonia—called ligands. A coordination complex focuses upon the metal arranging the ligands, while an inverse coordination complex focuses upon a non-metal atom/ion/organic molecule organising the metals.

An inverse coordination complex may or may not have internal bridging linkers and the bound metals may carry additional terminal ligands. In this reversed topology, the inverse coordination centre can be a monoatomic species which can bridge two or more metal atoms to stabilise a full range of familiar coordination geometries; for example, trigonal, tetrahedral, trigonal-bipyramidal, and octahedral.

Further, the inverse coordination centre can be a small polyatomic group. Typically, the central atom/polyatomic residue will feature nitrogen or other pnictogens (phosphorus, arsenic, antimony), oxygen or other chalcogens (sulphur, selenium, tellurium), and halogens (fluorine, chlorine, bromine, iodine). As an example, a "naked" phosphorus can function as a coordination centre, as can polyatomic phosphorus-containing species ranging from P_2, P_3...P_n to yield to all sorts of molecular architectures.

This book is dedicated to highlighting the fascinating array of inverse coordination complexes in a systematic manner by categorising the different classes with illustrative examples, in lieu of a comprehensive overview of this largely overlooked field. This book is intended to be a timely introduction to the field and as an essential instructional resource.

About the Authors

Ionel Haiduc is Professor Emeritus at Babes-Bolyai University in Cluj-Napoca, Romania. A distinguished chemist and prolific researcher, he has published over 340 journal articles and authored several books including *The Chemistry of Inorganic Ring Systems* (1970), *Basic Organometallic Chemistry* (1985), *The Chemistry of Inorganic Homo- and Heterocycles* (1987), *Organometallics in Cancer Chemotherapy* (1990), and *Supramolecular Organometallic Chemistry* (1999). He is a member of the Romanian Academy (former President from 2006 to 2014) and the Academia Europea (London), as well as an honorary or correspondent member of Göttingen Academy of Sciences and Humanities, Hungarian Academy of Sciences, Academy of Sciences of Moldova, and Montenegrin Academy of Sciences.

Edward R.T. Tiekink is Distinguished Professor and Head of the Research Centre for Crystalline Materials at Sunway University, Malaysia. He is a graduate of the University of Melbourne, Australia (D.Sc. 2006), where his passion for structural chemistry was nurtured. He has published in excess of 2,000 research papers/reviews and co-edited a number of books reflecting his interests in crystallography and metal-based drugs, including *Metallotherapeutic Drugs and Metal-Based Diagnostic Agents: The Use of Metals in Medicine* (2005), *Frontiers in Crystal Engineering* (2006), *Organic Crystal Engineering—Frontiers in Crystal Engineering* (2010), and *Multi-Component Crystals: Synthesis, Concepts, Function* (2017). He is a Fellow of the Royal Society of Chemistry and the Royal Australian Chemical Institute.

1 INTRODUCTION

1.1 Defining the concept

Coordination chemistry is a major chapter in inorganic chemistry that deals with "metal compounds (complexes) in which a central metal atom or ion as electron acceptor (Lewis acid) is surrounded by a number of electron donor ions or molecules (Lewis bases) defined as ligands" [1]. According to this definition, the metal atom plays the central role.

A reversed topology, however, is also possible in which a non-metal atom, ion, or even a small molecule can be regarded as the central moiety surrounded by a number of metal atoms, thus resulting in "inverse coordination" (Figure 1.1). A large number of such compounds are known today and this book aims to introduce and illustrate the novel chemical concept of inverse coordination.

A prototype of inverse coordination complexes is the triangular structure shown in Figure 1.2, but a much broader diversity exists. The internal bridging linkers can be one atom, two or more atom groups, or even absent (as will be illustrated in the following pages). The central atom (E) can be any of the pnictogens (nitrogen, phosphorus, arsenic, antimony), chalcogens (oxygen, sulphur, selenium, tellurium), or halogens (fluorine, chlorine, bromine, iodine) as naked atoms or as small polyatomic groups containing these non-metals. The topology of the structure shown in Figure 1.2 is described by the formulation $[(\mu_3\text{-}E)M_3(\mu_2\text{-}X)_3L_3]$, where E is the inverse coordination centre, M is the metal (transition metal, main group element, lanthanide, etc.), X is the internal bridging linker, and L is the peripheral (terminal) ligand.

1.2 Knowing the history

The history of inverse coordination complexes begins with trinuclear basic carboxylates. As detailed in a book chapter by Cannon and White [2], it was noted that in 1908, A. Werner and

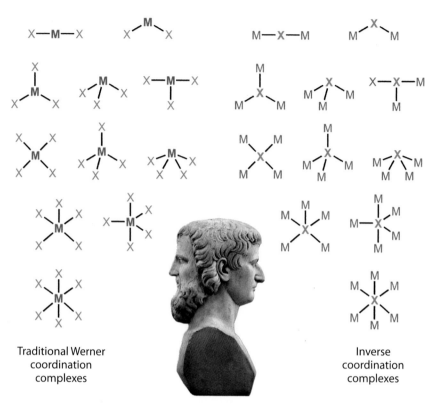

Traditional Werner
coordination
complexes

Inverse
coordination
complexes

Figure 1.1 The Janus relationship between traditional Werner complexes and inverse coordination complexes (Janus is the two-faced Roman god of beginnings and transitions, the god of change and of time; image of Janus from Svetlana Pasechnaya/Shutterstock.com)

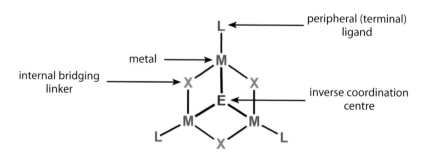

peripheral (terminal)
ligand

metal

internal bridging
linker

inverse coordination
centre

Figure 1.2 General prototype of inverse coordination complexes

R.F. Weinland independently found series of chromium compounds containing a group of three metal atoms and six carboxylates which remained intact in a variety of exchange reactions. This did not fit into Werner's coordination theory and could not be explained at the time. Only in 1960 did Orgel speculatively suggested the oxo-centred triangular structure [3], which was later confirmed for the chromium compound $[(\mu_3\text{-O})Cr_3(\mu_2\text{-OOCMe})_6(H_2O)_3]^+$ (**1**) through X-ray diffraction (Figure 1.3) [4,5]. Note that in Figure 1.3 and in subsequent chemical structure diagrams, the carbon atoms of the peripheral groups are shown in stick form and non-acidic hydrogen atoms are omitted.

After the technique of X-ray diffraction became widely available, numerous triangular oxo-centred complexes were structurally characterised and today, this family of compounds is well-represented [6]. Basic beryllium acetate was found to be a tetranuclear oxo-centred complex, $[(\mu_4\text{-O})Be_4(\mu_2\text{-OOCMe})_6]$ (**2**), and its solid-state structure (Figure 1.4) was first established in 1923 through X-ray crystallography by (Nobel laureate) Bragg and Morgan [7] and later confirmed by other research groups [8]. The compound was dubbed as an "inverse basic beryllium carboxylate complex" in an analytical chemistry journal [9], but inorganic chemists had not yet caught on to the idea of inverse coordination and the compound was presented in inorganic chemistry books of the day merely as a curiosity.

It was Mulvey who first used "inverse crown" archetype to describe cyclic structures in which "their arrangement of Lewis acidic [acceptor] and Lewis basic [donor] sites is opposite to that encountered in conventional crown ether complexes" [10]. With colleagues [11], he further noted that "they are 'inverse' crowns in the sense that their Lewis acidic/Lewis basic sites have been interchanged relative to those in conventional crown ether complexes, that is, here the metal atoms belong to the ring and not to the core" (Figure 1.5). The inexorable next step was the introduction of the term "inverse coordination", which was promoted by Mulvey and Wright [12].

The novel chemical concept of inverse coordination can be defined as "the formation of metal complexes in which the arrangement of acceptor and donor sites is opposite to that occurring in conventional coordination complexes" [13]. Note

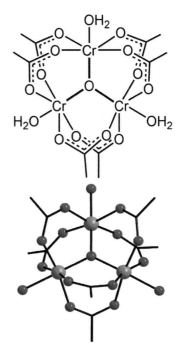

Figure 1.3 The chemical diagram and molecular structure of (**1**)

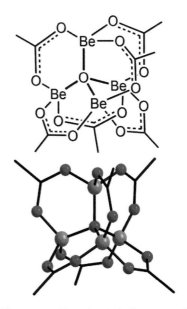

Figure 1.4 The chemical diagram and molecular structure of (**2**)

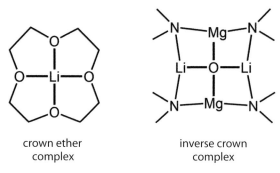

crown ether
complex

inverse crown
complex

Figure 1.5 Traditional and inverse crown complexes

that the terms "inverse metallacrown" and "metallacrown
complexes" [14,15] as well as "charge-reversed crown host"
[16], "anti-crown" [17], and—in the context of supramolecular
chemistry—"host-guest complexes" [18] are used alternatively for
this class of compounds.

In a 1986 review by Herrmann concerning "multiple bonds
between transition metals and 'bare' main group elements" [19],
non-metal coordination centres were recognised as inverse centres.
Some "chalcogenide-centred gold complexes" were also described
[20]. While not defined as inverse coordination complexes, the
major structure types were systematically analysed by Dance [21]
and very briefly reviewed [22]. Recently published comprehensive
reviews covering the inverse coordination complexes centred on
oxygen and other chalcogens [6], halogens [23], nitrogen [13], and
nitrogen-heteroatom molecules [24].

As will be amply demonstrated in the ensuing chapters,
the concept of inverse coordination complexes covers a vastly
greater range of chemical compounds than the crown complexes
exemplified in Figure 1.5. Thus, inverse coordination complexes
will be described with the central core being defined or occupied
by non-metal species, such as a single atom, ion, or small
molecule. Other variations do exist, such as when the central
core is surrounded by a number of metal atoms (ions) which may
be interconnected (or not) by internal bridging (intramolecular)
linkers, or when the coordinating metal atoms carry peripheral
(terminal) ligands to satisfy their preferred coordination number(s).

2 MONOATOMIC INVERSE COORDINATION CENTRES

2.1 Triangular complexes

2.1.1 Three-atom bridging linkers

As mentioned in the Introduction, the most numerous oxo-centred trimetallic inverse coordination complexes are those of the general formula $[(\mu_3\text{-}O)M_3(\mu_2\text{-}OOCR)_6L_3]$, often described as "basic carboxylates" [25]. The common feature of these complexes is that the central oxygen atom is connected to three metal atoms to form a $[(\mu_3\text{-}O)M_3]$ core with, or close to, a planar triangular geometry. The structure is stabilised by three pairs of bridging carboxylato linkers (triatomic internal linker bridges) attached to adjacent metal atoms and has additional peripheral (terminal) ligands L to satisfy the preferred coordination number of the specific metal atom.

A systematic evaluation of the chemical literature and of the Cambridge Structural Database (CSD) [26] revealed that several hundreds of such complexes have been characterised, usually through X-ray crystallography. Many have also been investigated for their physical properties; for example, magnetic behaviour in the case of some transition metals. For the sake of brevity, only selected examples will be cited in this book.

The known μ_3-oxo-centred trimetallic inverse coordination complexes reported so far, with six carboxylato internal linkers and of the general formula $[(\mu_3\text{-}O)M_3(\mu_2\text{-}OOCR)_6L_3]$ (3), contain both transition metals (including titanium, vanadium, chromium, molybdenum, tungsten, manganese, iron, ruthenium, cobalt, rhodium, and iridium) and main group elements (including aluminium and gallium) (Figure 2.1). For a recent review of these structures, refer to [6].

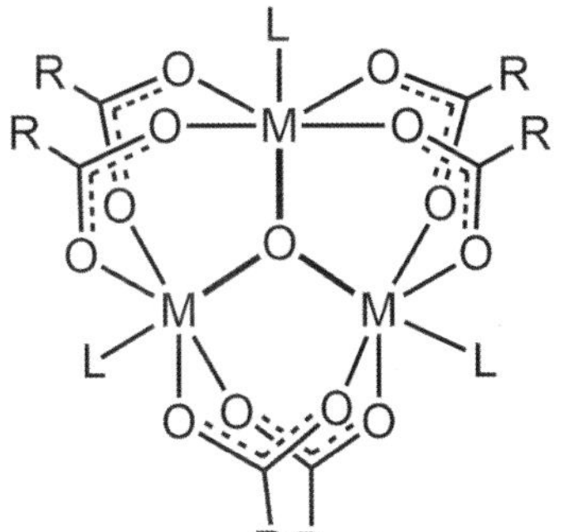

A broad diversity of triangular inverse coordination complexes can be formed by varying the metal M in the $[(\mu_3\text{-}O)M_3]$ core and the R group of the carboxylato internal bridging linkers

M = Ti, V, Cr, Mo, W, Mn, Fe, Ru, Co, Rh, Ir, Al and Ga

R = Me, Et, Ph, CG$_3$, etc.

L = H$_2$O, pyridine, THF, DMSO, etc.

Figure 2.1 The chemical diagram of (**3**)

5

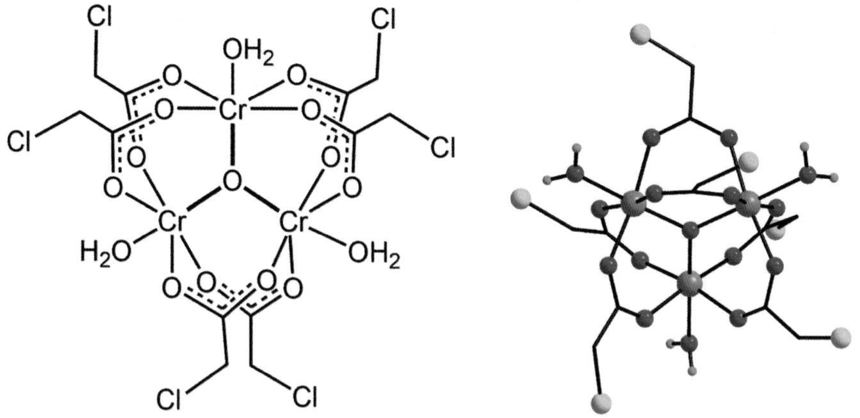

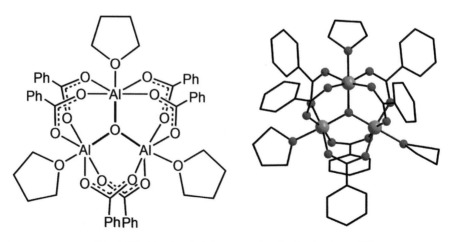

Figure 2.2 The chemical diagram and molecular structure of (**4**)

(or by replacing the carboxylates with other bidentate groups), and by using a variety of peripheral (terminal) L ligands. These possibilities are illustrated by the basic chromium formate $[(\mu_3\text{-O})\text{Cr}_3(\mu_2\text{-OOCH})_6(\text{H}_2\text{O})_3]^+$ [27], basic chromium benzoate $[(\mu_3\text{-O})\text{Cr}_3(\mu_2\text{-OOCPh})_6(\text{Py})_3]^+$ [28], basic chromium chloroacetate $[(\mu_3\text{-O})\text{Cr}_3(\mu_2\text{-OOCH}_2\text{Cl})_6(\text{OH}_2)_3]^+$ (**4**; Figure 2.2) [29], basic iron acetate $[(\mu_3\text{-O})\text{Fe}_3(\mu_2\text{-OOCMe})_6(\text{H}_2\text{O})_3]^+$ [30], and aluminium basic benzoate $[(\mu_3\text{-O})\text{Al}_3(\mu_2\text{-OOCPh})_6(\text{THF})_3]^+$ (**5**; Figure 2.3) [31]; note that Py is pyridine and THF is tetrahydrofuran. In several of these structures, metal⋯metal bonding interactions of varying strength are present but, for reasons of clarity, are not indicated.

Figure 2.3 The chemical diagram and molecular structure of (**5**)

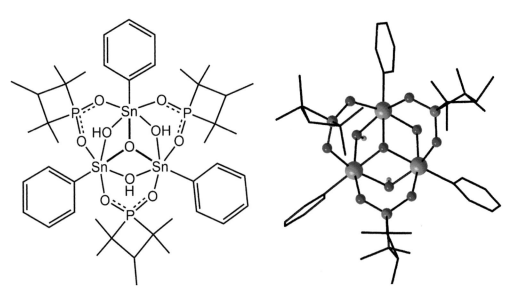

Figure 2.4 The chemical diagram and molecular structure of (**6**)

Many different types of oxo-acid anions are expected to function as internal bridging linkers in the same way that carboxylato anions function and yet, such examples are relatively rare. One such example is illustrated by the iron complex anion $[(\mu_3\text{-}O)Fe_3(\mu_2\text{-}SO_4)_6]^{5-}$ with sulphate as the internal bridging linker [32]. Another example with sulphate as the internal bridging linker is an exceptional nitrogen-centred, planar inorganic anion $[(\mu_3\text{-}N)Ir_3(\mu_2\text{-}SO_4)_6(H_2O)_3]^{4-}$ [33]. There are several complexes with organophosphorus anions providing internal bridging linkers leading to triangular inverse coordination complexes, with an illustrative example being $[(\mu_3\text{-}O)(PhSn)_3(\mu_2\text{-}OH)_3(\mu_2\text{-}O_2PC_3Me_5)_3]^+$ (**6**; Figure 2.4) [34].

Since the F^- anion is isoelectronic with the O^{2-} anion, some similar behaviour as an inverse coordination centre can be expected. This is demonstrated by the structure-directing role of fluoride in numerous organometallic fluorides [35]. While fluoro-centred complexes such as $[(\mu_3\text{-}F)M_3(\mu_2\text{-}OOCR)_6L_3]$ (**7**) are known, there are not many (Figure 2.5).

Some examples are the complexes with trifluoroacetate as the internal bridging linker, that is $[(\mu_3\text{-}F)M_3(\mu_2\text{-}OOCCF_3)_6L_3]^{n-}$ with M = Co and Ni, L = CF_3COOH, n = 1 [36], M = Mg, L = OMe, and Py, n = 3 [37], and the heterotrimetallic complexes

Figure 2.5 The chemical diagram of (**7**)

$[(\mu_3\text{-}F)Ni_2Cr(\mu_2\text{-}OOCBu^t)_6(L)_3]$ with L = Bu^{t}COOH, Py, and 4-MePy [38]. Another fluoride-centred, trimetallic inverse coordination complex is the copper complex $[(\mu_3\text{-}F)Cu_3(\mu_2\text{-}(Bu^t)_2$ PCH$_2$P(Bu^t)$_2$)$_3$]$^{2+}$ (8; Figure 2.6) [39] which also has an oxygen-centred, that is hydroxyl, analogue [40].

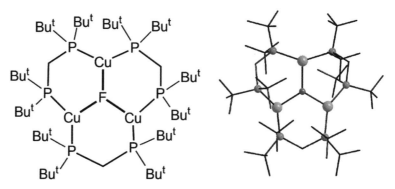

Figure 2.6 The chemical diagram and molecular structure of (**8**)

2.1.2 Two-atom bridging linkers

Two-atom bridging internal linkers are illustrated by some rare, oxo-centred, dicationic triplatinum inverse coordination complexes $[(\mu_3\text{-}OH)Pt_3(\mu_2\text{-}OPPh_2)_3(PPh_2Me)_3]^{2+}$ (9; Figure 2.7) [41] and dianionic $[(\mu_3\text{-}O)Pt_3(\mu_2\text{-}ONO)_3(NO_2)_3]^{2-}$ [42,43].

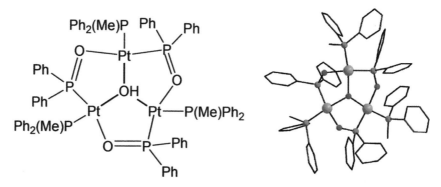

Figure 2.7 The chemical diagram and molecular structure of (**9**)

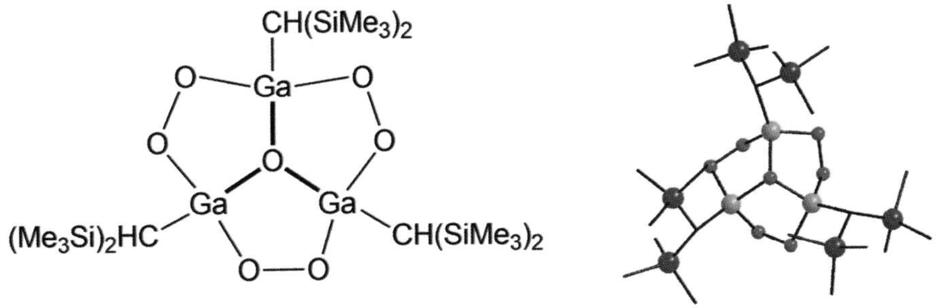

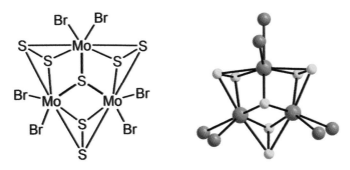

Figure 2.8 The chemical diagram and molecular structure of (**10**)

A unique peroxo organogallium compound $[(\mu_3\text{-}O)\{CH(SiMe_3)_3$ $Ga\}_3(\mu_2\text{-}OO)_3]$ (**10**; Figure 2.8) falls in the same category [44]. A sulphur analogue, that is dianionic $[(\mu_3\text{-}S)Mo_3(\mu_2\text{-}S_2)_3Br_6]^{2-}$ (**11**; Figure 2.9) [45], also falls in this category. While there are only two bonds to the O_2 internal bridging linker in compound (**10**), there are four points of connection to the S_2 linker in compound (**11**), differences correlated with the different orientations of O_2 versus S_2.

An important class of μ_3-oxo-centred trimetallic inverse coordination complexes with two-atom internal bridging linkers are compounds in which the internal bridging linkers are species containing nitrogen donor atoms, of which a prominent example is the one comprising nitrogen heterocycles. In this context, there are many inverse coordination complexes with pyrazolato anions (Hpz = pyrazole) functioning as the internal bridging linkers [45–47]. Complexes featuring six pyrazole rings as internal

Figure 2.9 The chemical diagram and molecular structure of (**11**)

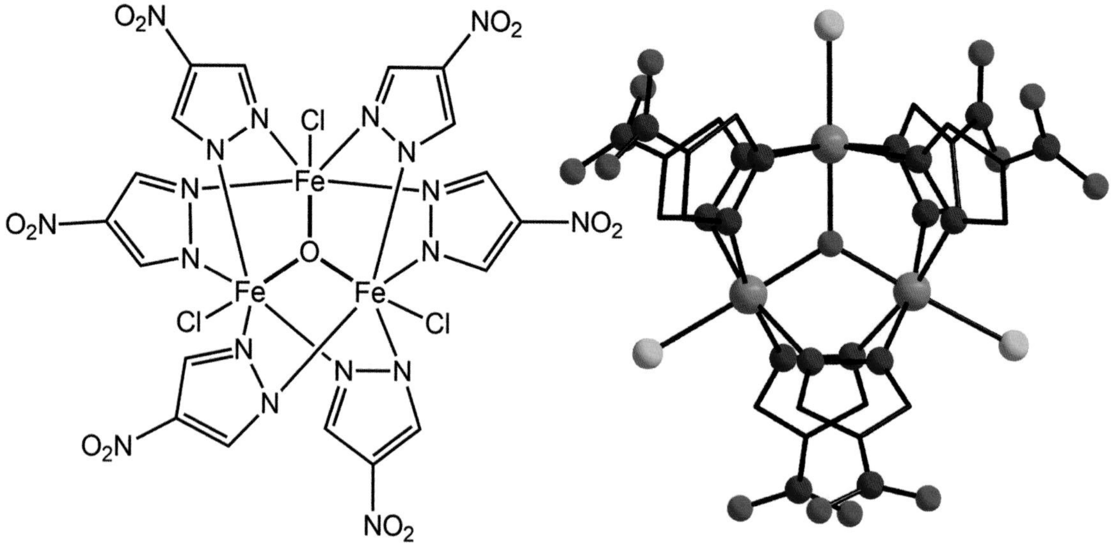

Figure 2.10   The chemical diagram and molecular structure of (**12**)

bridging linkers are illustrated by the iron complexes $[(\mu_3\text{-}O)Fe_3(\mu_2\text{-}4\text{-}NO_2\text{-}pz)_6X_3]^{2-}$, with X = Cl (**12**; Figure 2.10) and Br [48].

In the same way, a rather large number of inverse coordination complexes with three pyrazole rings as internal bridging linkers are known. These include $[(\mu_3\text{-}O)Cu_3(\mu_2\text{-}pz)_3Cl_3]^{2-}$ (**13**; Figure 2.11) [49] and compounds with substituted pyrazolyl $[(\mu_3\text{-}O)Cu_3(\mu_2\text{-}4\text{-}X\text{-}pz)_3X_3]^{2-}$ (X = H, Cl, Br, NO$_2$, CHO) [50].

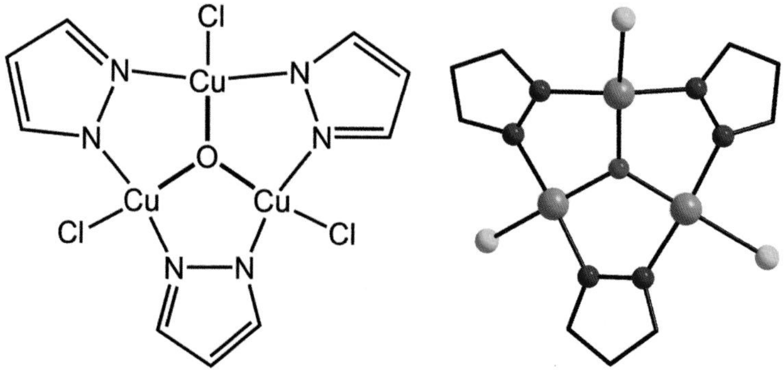

Figure 2.11   The chemical diagram and molecular structure of (**13**)

2.1.3 Single-atom bridging linkers

In inverse coordination complexes with various central cores, $[(\mu_3\text{-E})M_3(\mu_2\text{-X})_3]$, single-atom internal bridging linkers are usually oxygen, halogen, and sulphur atoms or parts of hydroxo, alkoxo, and thiolato bridges. Several oxygen-centred complexes in this family are known. Among these are the cyclopentadienyl compounds of zirconium, $[(\mu_3\text{-O})\{Zr(Cp)_2\}_3(\mu_2\text{-O})_3]$ (**14**; Figure 2.12) [51], and molybdenum, $[(\mu_3\text{-O})Mo_3(\mu_2\text{-O})_3(H_2O)_9]^{4+}$ (**15**; Figure 2.13) [52]; Cp is $\eta^5\text{-C}_5H_5$, that is, the cyclopentadienyl monoanion.

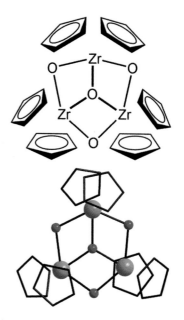

Figure 2.12   The chemical diagram and molecular structure of (**14**)

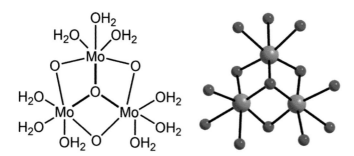

Figure 2.13   The chemical diagram and molecular structure of (**15**) (note that water-hydrogen atoms are not shown)

Sulphur internal linkers occur in an iron complex with thiolato bridging linkers $[(\mu_3\text{-O})Fe_3(\mu_2\text{-SPh})_3\{N(SiMe_3)_2\}_3]$ (**16**; Figure 2.14) [53]. Sulphur-centred inverse coordination complexes with single-atom internal bridging linkers are well known; for example, tetracationic $[(\mu_3\text{-S})Mo_3(\mu_2\text{-S})_3(H_2O)_6]^{4+}$ [54] and monoanionic $[(\mu_3\text{-S})Ni_3(\mu_2\text{-SBu}^t)_3(CN)_3]^-$ (**17**; Figure 2.15) [55].

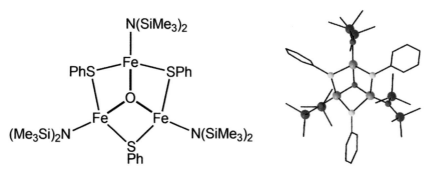

Figure 2.14   The chemical diagram and molecular structure of (**16**)

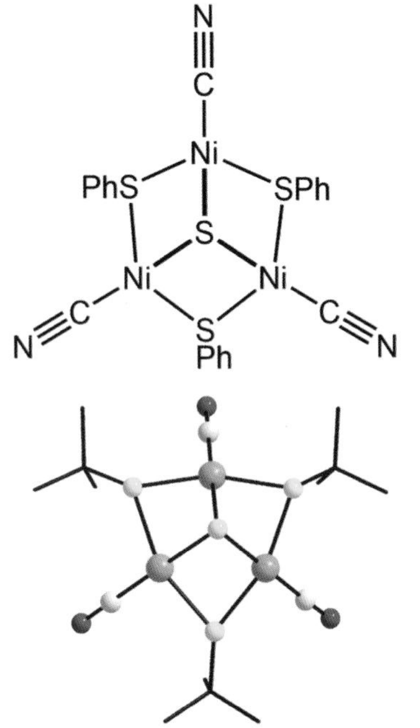

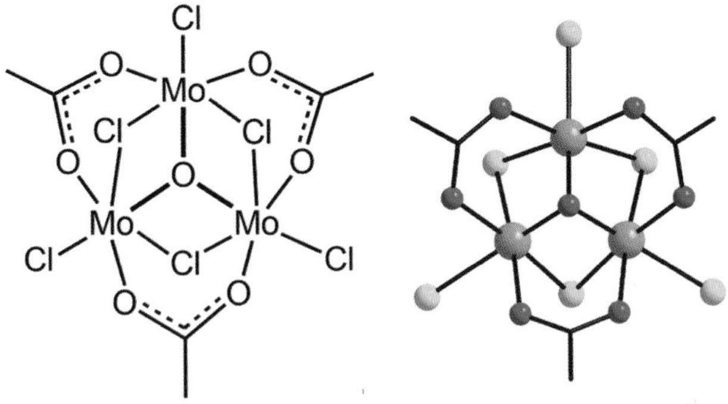

Figure 2.16   The chemical diagram and molecular structure of (**18**)

Figure 2.15   The chemical diagram and molecular structure of (**17**)

Halogen internal bridging linkers were revealed in the structures of $[(\mu_3\text{-O})\text{Mo}_3(\mu_2\text{-X})_3(\mu_2\text{-OOCMe})_3\text{X}_3]^-$, where X = Cl (**18**; Figure 2.16) and Br [56]; $[(\mu_3\text{-N})(\text{SnMe}_2)_3(\mu_2\text{-X})_3]$ where X = Cl, Br, I (**19**; Figure 2.17) [57]; and $[(\mu_3\text{-P})\text{Bi}_3(\mu_2\text{-Cl})_3\{\text{CH}(\text{SiMe}_3)_2\}_3]$ (**20**; Figure 2.18) [58].

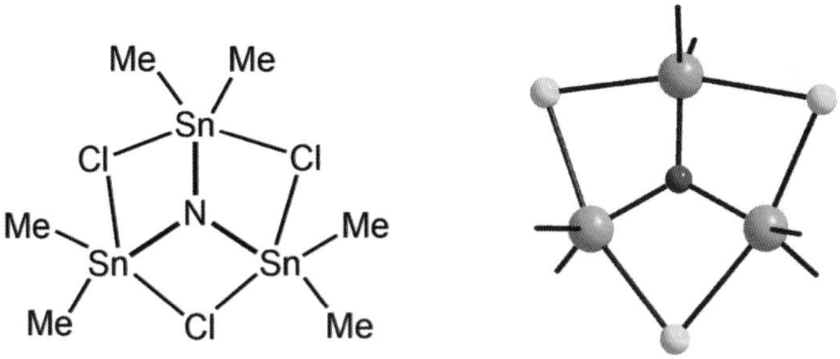

Figure 2.17   The chemical diagram and molecular structure of (**19**)

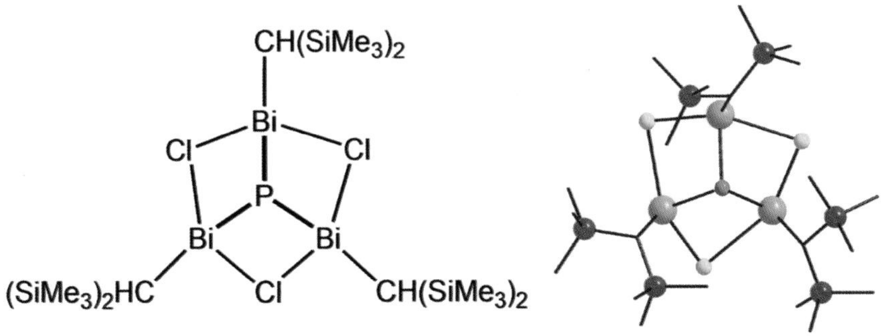

Figure 2.18   The chemical diagram and molecular structure of (**20**)

Internal bridging linkers based on nitrogen are less common but can be illustrated with the nitrogen-centred titanium inverse coordination complex [(μ₃-N)(TiCp*)₃(μ₂-NH)₃] (**21**; Figure 2.19) [59].

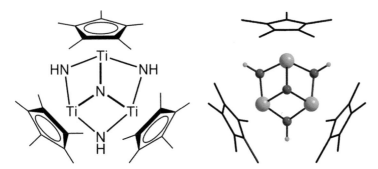

Figure 2.19   The chemical diagram and molecular structure of (**21**)

2.2 Tetrahedral complexes

2.2.1 Three-atom bridging linkers

Particularly notable in this category is the "basic beryllium acetate" structure, [(μ₄-O)Be₄(μ₂-OOCMe)₆] (Figure 1.4), which has been known for a long time—its structure attracted early researchers as it was among the first identified as an inverse coordination complex [7]. A tetrahedral motif related to "basic beryllium acetate" has numerous examples, including basic zinc carboxylates as illustrated by [(μ₄-O)Zn₄(μ₂-OOCPh)₆] (**22**; Figure 2.20) [60].

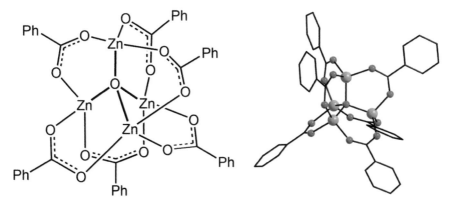

Figure 2.20   The chemical diagram and molecular structure of (**22**)

Organophosphorus anions can also serve as internal bridging linkers, and several inverse coordination complexes are known to have alkylphosphates, such as [(μ4-O)Zn4{O2P(OBu^t)2}6] [61], and dithiophosphates, such as [(μ4-O)Zn4{μ2-S2P(OBu^t)2}6] [62]. Similar inverse coordination complexes have been structurally characterised by X-ray crystallography with monothiophosphates, such as [(μ4-O)Zn4{μ2-SOP(OPh)2}6] (23; Figure 2.21) [62], and dithiophosphinates [(μ4-O)Zn4{μ2-S2PEt2}6] (24; Figure 2.22) [63]. Sulphur-centred inverse coordination complexes with dithiophosphate [64] and even dithioarsinate, such as

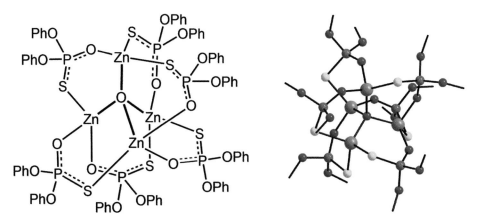

Figure 2.21 The chemical diagram and molecular structure of (**23**) (note that only ipso-C atoms of the phenyl rings are shown)

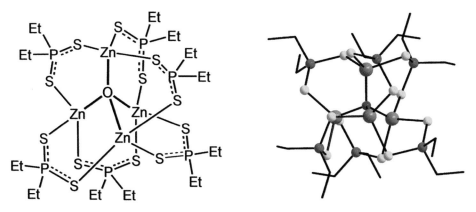

Figure 2.22 The chemical diagram and molecular structure of (**24**)

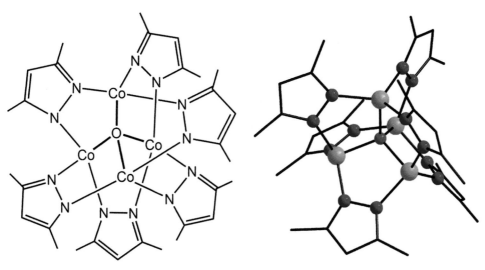

Figure 2.23 The chemical diagram and molecular structure of (**25**)

$[(\mu_4\text{-}S)Zn_4(\mu_2\text{-}S_2AsMe_2)_6]$ [65], internal bridging linkers have also been described. Finally, an oxo-centred, tetrahedral inverse coordination copper complex with an intricate bis(pyrazolyl) metane bridging linker was surprisingly formed [66].

2.2.2 Two-atom bridging linkers

The N–N bond of pyrazole- and triazole-based heterocycles can serve as two-atom bridging linkers in tetrahedral oxo-centred inverse coordination complexes. However, this topology is comparatively rare; an example is illustrated by $[(\mu_4\text{-}O)Co_4(\mu_2\text{-}dmpz)_6]$ (Hdmtz = 2,4-dimethylpyrazole) (**25**; Figure 2.23) [67].

2.2.3 Single-atom bridging linkers

There are numerous inverse coordination complexes assembled via single-atom bridging linkers, like the μ_4-oxo-centred inorganic adamantanes. Archetypical members of this class are a series of approximately 40 copper complexes of the general formula $[(\mu_4\text{-}O)Cu_4(\mu_2\text{-}X)_6L_4]$ which have been thoroughly analysed in a recent review [68]; further examples have subsequently been published. Similar structures are known for other transition metals, with that of a chromium species, $[(\mu_4\text{-}O)Cr_4(\mu_2\text{-}Cl)_6(THF)_4]$ (**26**; Figure 2.24) [69], being a good exemplar.

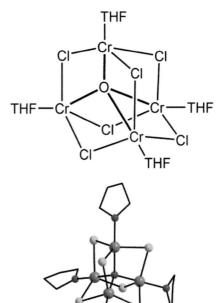

Figure 2.24 The chemical diagram and molecular structure of (**26**)

A rare topology based upon tetrahedral inverse coordination centres is evident in a pair of gallium- and indium-spiro complexes $[(\mu_4\text{-}S)(\mu_2\text{-}MMe_2)_6(\mu_2\text{-}SSiMe_3)_4]$ for M = Ga and In (**27**; Figure 2.25) [70].

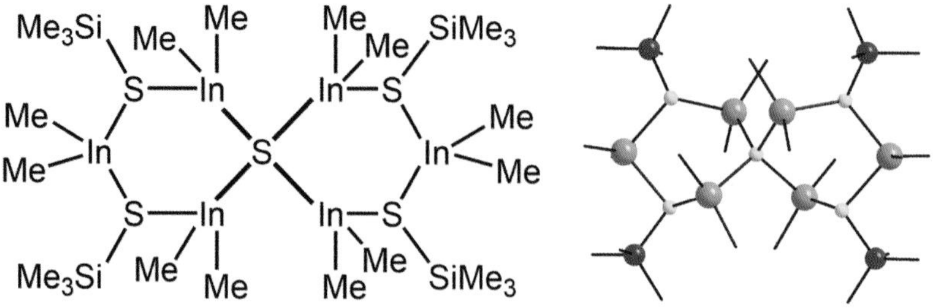

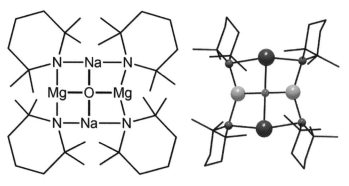

Figure 2.25 The chemical diagram and molecular structure of (**27**)

2.3 Planar complexes

A surprising result of tetra-connective oxygen functioning as an inverse coordination centre is the adoption of a square-planar geometry for the oxygen centre. This is achieved in groups of complexes of the type $[(\mu_4\text{-}O)M_2M'_2(\mu_2\text{-}\{N(SiMe_3)_2\})_4]$, where M = Mg and M' = Li, Na; M = Zn, Mn and M' = Na, K; M = Co and M' = Na as exemplified by compound (**28**) where M = Zn and M' = Na (Figure 2.26) [71]; as well as $[(\mu_4\text{-}O)M_2M'_2(tmp)_4]$ (tmp = tetramethylpiperidine), where M = Mg, Mn and M' = Li, Na as illustrated by compound (**29**) where M = Mg and M' = Na (Figure 2.27) [72].

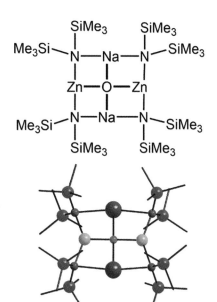

Figure 2.26 The chemical diagram and molecular structure of (**28**)

Figure 2.27 The chemical diagram and molecular structure of (**29**)

A special inverse coordination complex is the hydroxo-centred and completely inorganic anion $[(\mu_4\text{-OH})Cu_4(\mu_2\text{-CO}_3)_8]^{9-}$ with copper(I) and eight carbonato internal bridging linkers (**30**; Figure 2.28) [73].

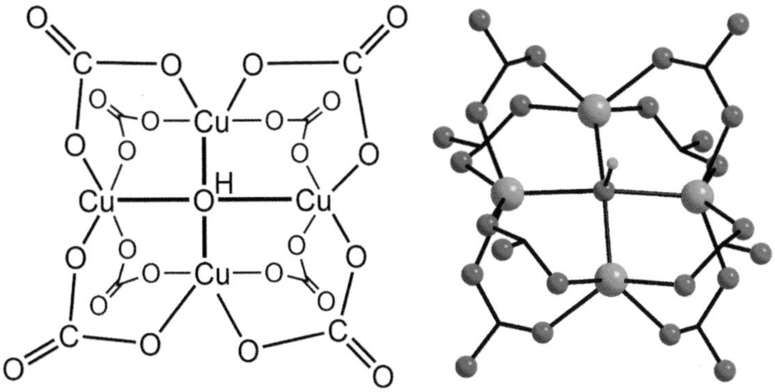

Figure 2.28   The chemical diagram and molecular structure of (**30**)

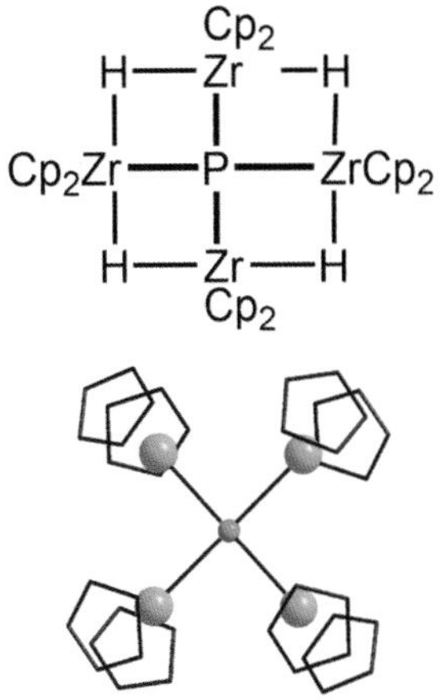

Figure 2.29   The chemical diagram and molecular structure of (**31**)

Phosphorus rarely displays a tetra-connective planar geometry but one exceptional case is illustrated by the cationic complex $[(\mu_4\text{-P})(ZrCp_2)_4(\mu_2\text{-H})_4]^+$ (**31**; Figure 2.29); the bridging hydride atoms were not located in the X-ray structure analysis [74].

Two exotic inverse coordination complexes of neodymium, namely anionic $[(\mu_4\text{-Cl})Nd_4(\mu_2\text{-SPh})_8\{\eta^1\text{-N(SiMe}_3)_2\}_4]^-$ (**32**; Figure 2.30) [75] and $[(\mu_4\text{-Cl})(NdCp^*)_4(\mu_2\text{-Cl})_8]^-$ (**33**; Figure 2.31) [76], are particularly noteworthy as each contains a planar tetra-connective chloride centre.

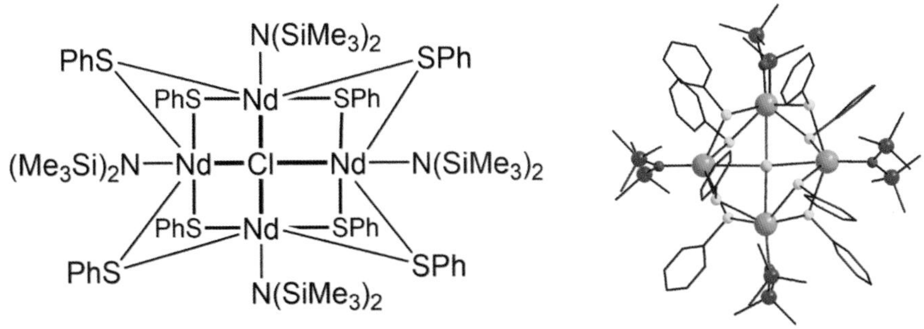

Figure 2.30   The chemical diagram and molecular structure of (**32**)

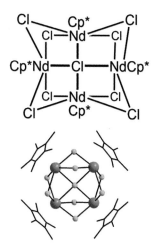

Figure 2.31 The chemical diagram and molecular structure of (**33**)

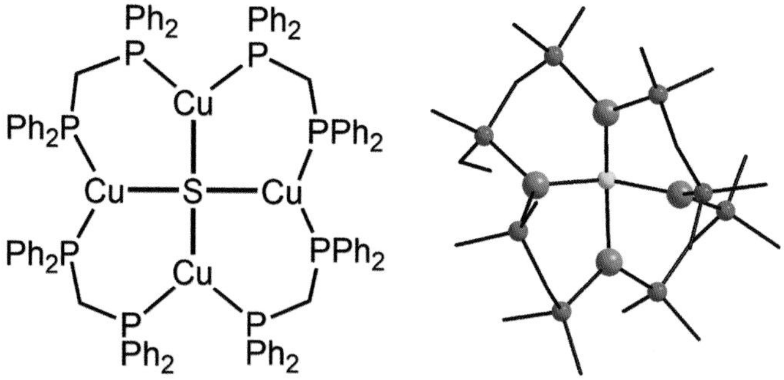

Figure 2.32 The chemical diagram and molecular structure of (**34**) (note that only ipso-carbon atoms of the phenyl groups are shown)

A sulphur-centred complex with a flattened square-pyramidal geometry, $[(\mu_4\text{-S})Cu_4(\mu_2\text{-Ph}_2PCH_2PPh_2)_4]^{2+}$ (**34**; Figure 2.32) [77], represents a new type of tetra-connective topology.

An unusual planar inverse coordination geometry is displayed by penta-connective halide atoms in some rare examples of alkali metal complexes. A spectacular $[(\mu_5\text{-Cl})M_5]$ inverse coordination core occurs in the anions $[(\mu_5\text{-X})M_5\{\mu_2\text{-N(SiMe}_3)_2\}_5]^-$ with X = Cl, Br and I, M = Li, Na (**35**; Figure 2.33), in which the central halogen is coplanar with the five alkali metal atoms [78].

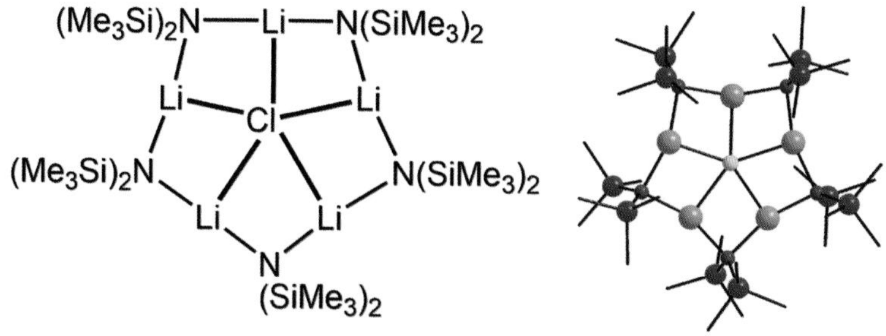

Figure 2.33 The chemical diagram and molecular structure of (**35**) (X = Cl and M = Li)

2.4 Trigonal-bipyramidal and square-pyramidal complexes

Penta-connective inverse coordination centres usually produce typical trigonal-bipyramidal and square-pyramidal topologies.

Figure 2.34 The chemical diagram and molecular structure of (**36**)

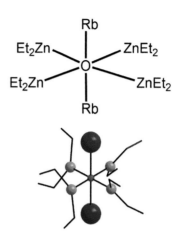

Unexpected outcomes from the reactions of diethylzinc with several alkali metal hydroxides gave rise to oxo-centred inverse coordination cores [79]. In the case of sodium hydroxide, a trigonal-bipyramidal geometry for oxygen was observed in [(μ_5-O)Na$_2$(ZnEt$_2$)$_3$] (**36**; Figure 2.34). When potassium and rubidium hydroxides were employed, the result was oxo-centred cores with octahedral hexa-connective oxygen, such as [(μ_6-O) M$_2$(ZnEt$_2$)$_3$] (M = K, Rb) (**37**; Figure 2.35). In both motifs, the zinc atoms define equatorial positions and the respective alkali metal occupies axial positions [79].

Figure 2.35 The chemical diagram and molecular structure of (**37**) (M = Rb)

A number of lanthanides are known to form μ_5-oxo-centred inverse coordination complexes. The lanthanide complex [(μ_5-O) Nb$_5$(μ_3-OPr^i)$_2$(μ_2-OPr^i)$_6$(OPr^i)$_5$(HOPr^i)$_2$] (**38**; Figure 2.36) [80] is an example of a μ_5-oxo-centred inverse coordination complex based on a trigonal-bipyramidal geometry. This is, in fact, a rare

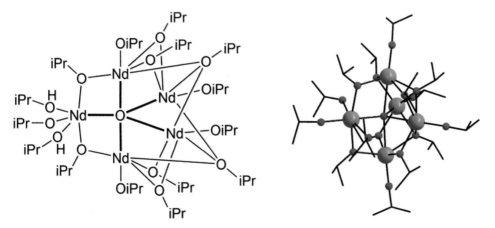

Figure 2.36 The chemical diagram and molecular structure of (**38**)

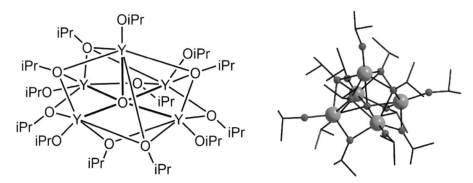

Figure 2.37 The chemical diagram and molecular structure of (**39**)

example as most species are constructed with square-pyramidal μ_5-oxo-centred cores as seen in the complex $[(\mu_5\text{-O})Y_5(\mu_3\text{-OPr}^i)_4(\mu_2\text{-OPr}^i)_4(\text{OPr}^i)_5$ (**39**; Figure 2.37) [81].

In the same way, a large series of iron compounds of the general formula $[(\mu_5\text{-O})Fe_5(\mu_2\text{-OR})_8X_5]$, with various combinations of X and R have been described. Each features a square-pyramidal geometry for the $[(\mu_5\text{-O})Fe_5]$ core; for example, X = Cl and R = i-Pr (**40**; Figure 2.38) [82].

Penta-connective fluorine was found in a lithium complex with N,N′-bis(trimethylsilyl)pentafluorobenzamidine internal bridging linkers, in the compound $[(\mu_5\text{-F})Li_5\{\mu_2\text{-C}_6F_5C(NSiMe_3)_2\}_4]$ (**41**; Figure 2.39) [83].

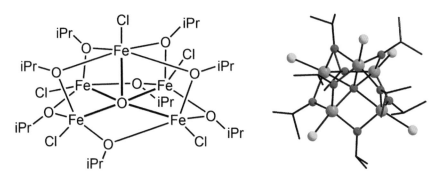

Figure 2.38 The chemical diagram and molecular structure of (**40**)

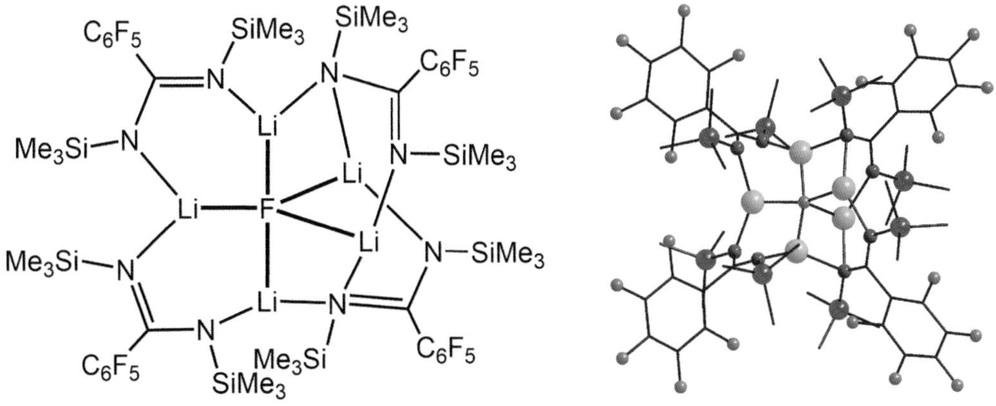

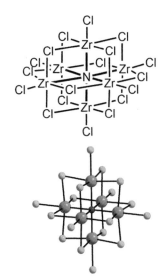

Figure 2.39   The chemical diagram and molecular structure of (**41**)

2.5 Octahedral complexes

There are a large number of inverse coordination complexes featuring hexa-connective oxygen within $[(\mu_6\text{-O})M_6]$ cores. Prominent among these is a series of lanthanide oxo-centred complex cations of the general formula $[(\mu_6\text{-O})Ln_6(\mu_3\text{-OH})_8 (H_2O)_{24}]^{8+}$, for Ln = Nd, Eu, Gd, Tb, and Dy, having eight μ_3-OH groups on the octahedral faces and four terminally bound water molecules on each lanthanide atom [84].

Polyoxometallates, also known as Lindquist structures, include hexa-vanadates, -tantalates, -molybdates, and -tungstates, and can also be classified as inversion coordination complexes—their structures feature $[(\mu_6\text{-O})M_6]$ cores connected to six metal atoms [85]. While they are of interest, these structures will not be discussed further as they have been extensively covered in the literature [86].

Zirconium forms an octahedral nitrogen-centred inverse coordination complex with chlorido internal bridging linkers in the complex $[(\mu_6\text{-N})Zr_6(\mu_2\text{-Cl})_{12}Cl_6]^{4-}$ (**42**; Figure 2.40) [87]. An example of an encapsulated fluoride anion as a μ_6-F core is found in the heterobimetallic complex $[(\mu_6\text{-F})\{Ti(\eta^5\text{-C}_5H_4SiMe_3)\}_5Al(\mu_2\text{-F})_{12} (THF)]$ (**43**; Figure 2.41) [88].

Figure 2.40   The chemical diagram and molecular structure of (**42**)

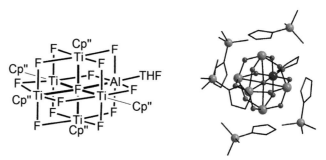

Figure 2.41   The chemical diagram and molecular structure of (**43**) (Cp" is η^5-C$_5$H$_4$SiMe$_3$)

As a final illustrative example, hexa-connective chloride features in the structure of $[(\mu_6\text{-Cl})\text{Ti}_6(\mu_2\text{-Cl})_{12}\{\text{N}(\text{C}_6\text{H}_3\text{-2,6-}(\text{Pr}^i)_2\}_6]^-$ (**44**; Figure 2.42) [89], which exhibits the same basic $\text{M}_6(\mu_2\text{-Cl})$ framework as complexes (**42**) and (**43**) respectively.

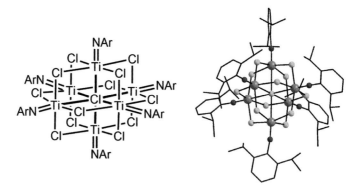

Figure 2.42   The chemical diagram and molecular structure of (**44**)

2.6 Linker-free complexes

In the previous sections, the presented inverse coordination complexes contain internal bridging linkers which supposedly stabilise the adopted topology. However, there are also numerous examples of inverse coordination complexes without bridging linkers, and which are formed just through interactions between the central atom and a requisite number of metal moieties. These are mostly trigonal and tetragonal complexes with the central atom displaying various geometries. In these examples, the central atoms can be oxygen, sulphur, nitrogen, phosphorus, or halogen.

Tri-connective centres with a trigonal-pyramidal environment are known in some oxygen- and sulphur-centred inverse coordination complexes of gold [(μ$_3$-O)Au$_3$(PPh$_3$)$_3$]$^+$ (**45**; Figure 2.43) [90], [(μ$_3$-S) Au$_3$(PPh$_3$)$_3$]$^+$ [91], and [(μ$_3$-S)Au$_3$(C$_6$F$_5$)$_3$]$^{2+}$ (**46**; Figure 2.44) [92].

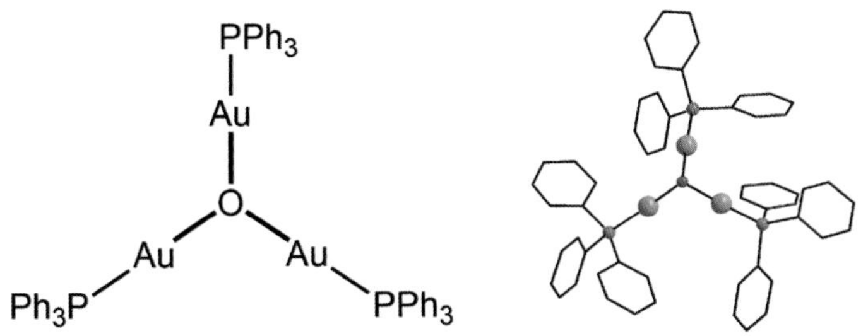

Figure 2.43 The chemical diagram and molecular structure of (**45**)

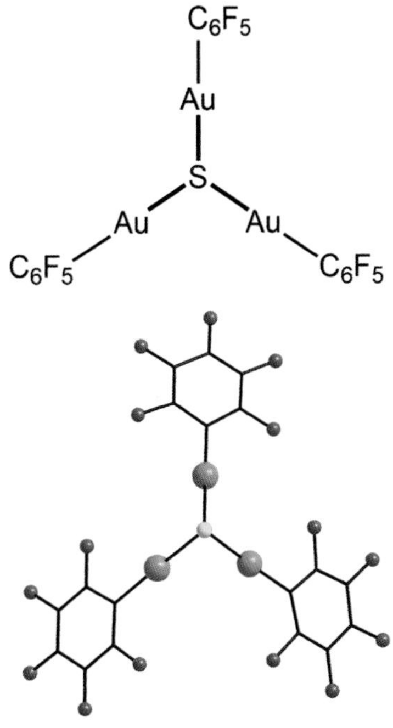

A number of trigonal-pyramidal inverse coordination complexes contain organometallic moieties, such as in organomercury [(μ$_3$-S) (HgMe)$_3$]$^+$ (**47**; Figure 2.45) [93] as well as organotin [(μ$_3$-P) (SnPh$_3$)$_3$] (**48**; Figure 2.46) [94] and organoarsenic [(μ$_3$-N) (AsMe$_2$)$_3$] (**49**; Figure 2.47) [95] compounds.

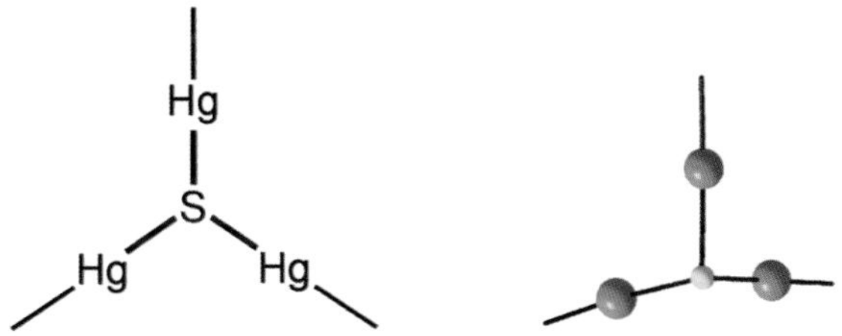

Figure 2.44 The chemical diagram and molecular structure of (**46**)

Figure 2.45 The chemical diagram and molecular structure of (**47**)

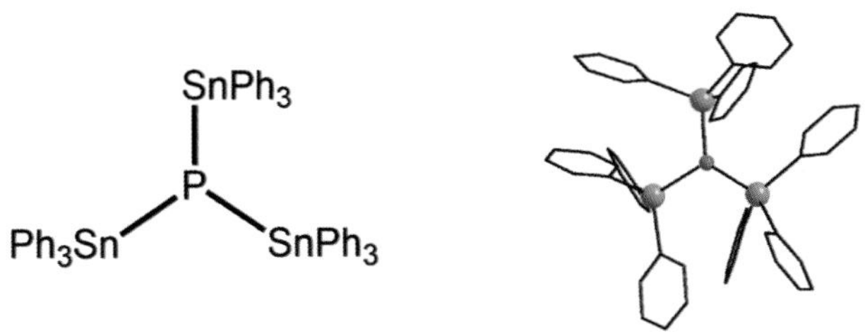

Figure 2.46 The chemical diagram and molecular structure of (**48**)

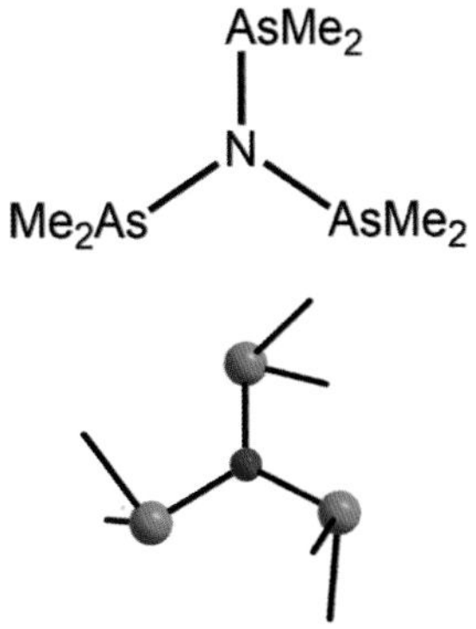

Figure 2.47 The chemical diagram and molecular structure of (**49**)

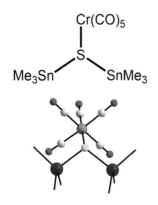

Figure 2.48 The chemical diagram and molecular structure of (**50**)

Unusual sulphur-centred heterobimetallic complexes with organometallic and metal carbonyl groups [(μ₃-S)(SnMe₃)₂Cr(CO)₅] (**50**; Figure 2.48) and [(μ₃-S)(PbMe₃)₂W(CO)₅] have also been described [96].

Some triangular inverse coordination complexes are planar. These include the sulphur-centred rhenium complex [(μ₃-S)Re₃(CO)₁₂]⁺ (**51**; Figure 2.49) [97], the organomercury compound [(μ₃-Br){Hg(C₆F₅)₃}]⁻ (**52**; Figure 2.50) [98], and the organoantimony compounds [(μ₃-O)(SbMe₂)₃]⁺ (**53**; Figure 2.51) [99] and [(μ₃-N)(SbMe₂)₃] (**54**; Figure 2.52) [100]. Linker-free tetragonal inverse coordination complexes are less frequent and display tetrahedral, square-pyramidal and planar geometries.

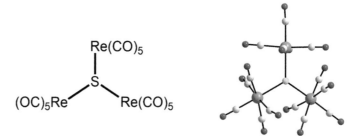

Figure 2.49 The chemical diagram and molecular structure of (**51**)

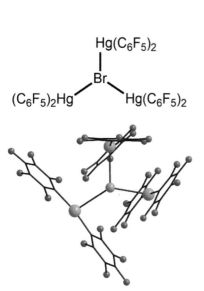

Figure 2.50 The chemical diagram and molecular structure of (**52**)

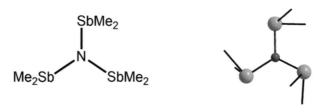

Figure 2.51 The chemical diagram and molecular structure of (**53**)

Figure 2.52 The chemical diagram and molecular structure of (**54**)

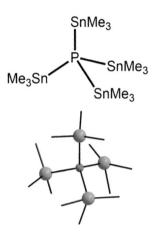

Figure 2.53 The chemical diagram and molecular structure of (**55**)

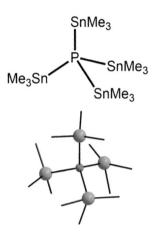

Figure 2.54 The chemical diagram and molecular structure of (**56**)

Tetrahedral inverse coordination complexes are nitrogen- and phosphorus-centred with gold donors, such as [(μ_4-N)Au$_4$(PPh$_3$)$_4$]$^+$ (**55**; Figure 2.53) [101] and [(μ_4-P)Au$_4$(PBu$^t_3$)$_4$]$^+$ [102], and organotin donors, such as [(μ_4-N)(SnMe$_3$)$_4$]$^+$ and [(μ_4-P)(SnMe$_3$)$_4$]$^+$ (**56**; Figure 2.54) [103].

The observation of a square-planar geometry in inverse coordination complexes is rare. One example is found in the sulphur-centred gold inverse coordination complex [(μ_3-S)Au$_4$(PPh$_3$)$_4$}]$^{2+}$ (**57**; Figure 2.55) [104].

Two illustrative examples of planar tetragonal inverse coordination complexes are found in the halogen-centred organoantimony compounds [(μ_4-Cl)(SbPh$_2$)$_4$(PMe$_3$)$_4$]$^{3+}$ (**58**; Figure 2.56) [105] and [(μ_4-I)(SbPh)$_4$I$_8$]$^-$ (**59**; Figure 2.57) [106].

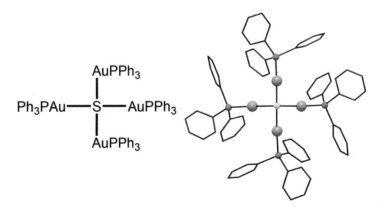

Figure 2.55 The chemical diagram and molecular structure of (**57**)

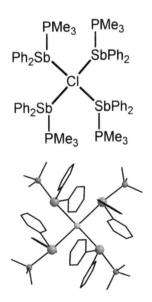

Figure 2.56 The chemical diagram and molecular structure of (**58**)

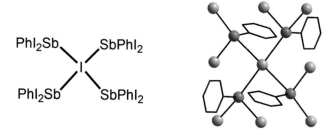

Figure 2.57 The chemical diagram and molecular structure of (**59**)

Penta-connective nitrogen forms a square-pyramidal gold inverse coordination complex $[(\mu_5\text{-N})Au_5(PPh_3)_5)]^{2+}$ (**60**; Figure 2.58) [107] while penta-connective phosphorus occurs in the square-pyramidal $[(\mu_5\text{-P})(AuPPh_3)_5]^{2+}$ cation (**61**; Figure 2.59) [108]. The existence of octahedral $[(\mu_6\text{-P})(AuL)_6]^{3+}$ (L = PBu$^t_3$, PPr$^i_3$) complexes containing hexa-connective phosphorus has been established with the aid of NMR and mass spectrometry [109].

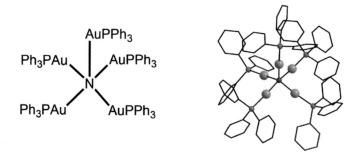

Figure 2.58 The chemical diagram and molecular structure of (**60**)

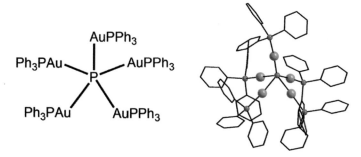

Figure 2.59 The chemical diagram and molecular structure of (**61**)

3 POLYATOMIC INVERSE COORDINATION CENTRES

3.1 Polynitrogen as the inverse coordination centre

3.1.1 Dinitrogen

Numerous inverse coordination complexes with dinitrogen coordination centres have been discovered and investigated during studies on nitrogen fixation and activation [110]. However, these were not generally described as inverse coordination complexes until quite recently [13], which is appropriate as the dinitrogen centres play a crucial structure-directing role and cannot be regarded simply as just bridging.

There are two types of inverse coordination complexes centred about dinitrogen: "head-on/head-on" and "side-on/side-on". In both types, the central dinitrogen group can maintain a triple bond character, or the bond order can be reduced to a double or even a single bond. These considerations are readily reflected in the nitrogen–nitrogen interatomic distances. The subject has attracted much interest and is covered in numerous reviews, including one with an emphasis on structural aspects [111].

A selection of examples illustrating the head-on/head-on inverse coordination with dinitrogen as the centre includes the complexes $[(\mu_2\text{-N}{\equiv}\text{N})(\text{ZrCp})_2\{\text{HC}(\text{SiMe}_3)_2\}_2]$ (**62**; Figure 3.1) [112], $[(\mu_2\text{-N}{=}\text{N})$ $\text{V}_2(\text{CH}_2\text{Bu}^t)_6]$ (**63**; Figure 3.2) [113], and $[(\mu_2\text{-N-N})(\text{MoMe}_3\text{Cp}^*)_2]$ (**64**; Figure 3.3) [114], featuring triply-, doubly- and singly-bonded dinitrogen respectively; Cp^* is $\eta^5\text{-C}_5\text{Me}_5$, that is the pentamethylated

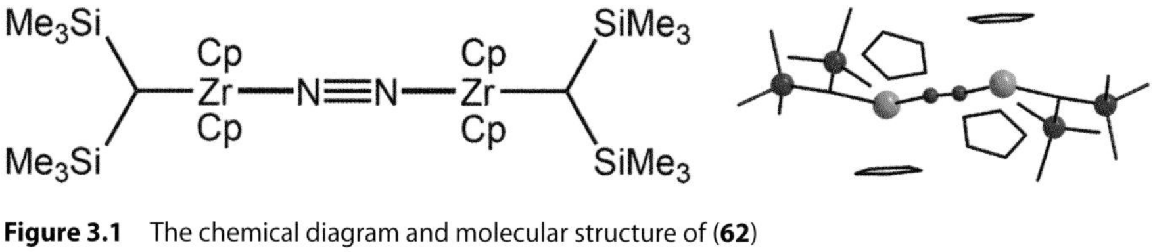

Figure 3.1 The chemical diagram and molecular structure of (**62**)

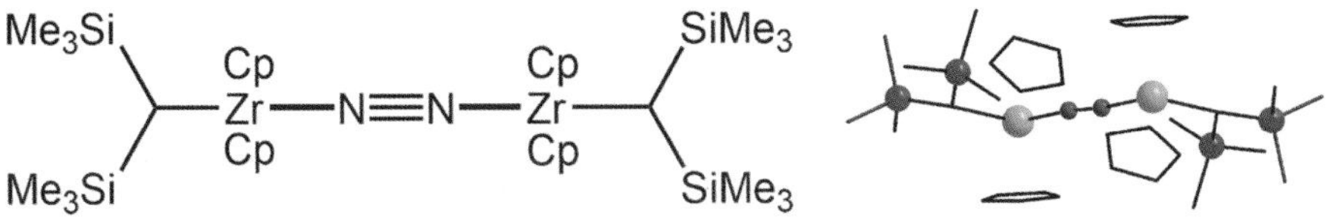

Figure 3.2   The chemical diagram and molecular structure of (**63**)

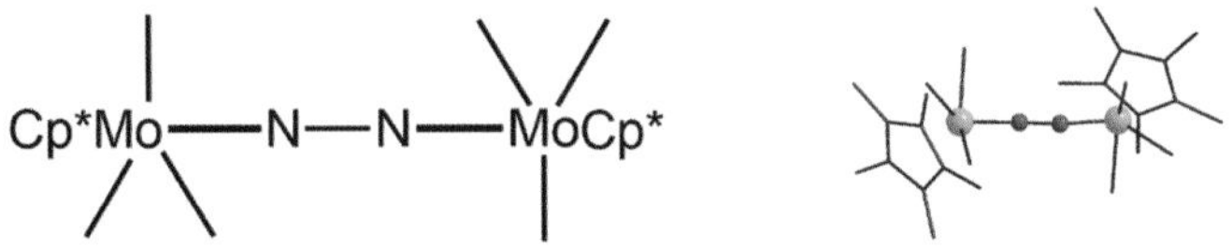

Figure 3.3   The chemical diagram and molecular structure of (**64**)

cyclopentadienyl monoanion. In all three molecules, the dinitrogen species connects two metals in an approximately linear fashion.

Side-on/side-on coordination is also displayed in numerous dinitrogen inverse coordination complexes and is illustrated by [(μ_2-N≡N){Zr(η^5-C$_5$H$_3$(SiMe$_3$)$_2$}$_2$] (**65**; Figure 3.4) [115].

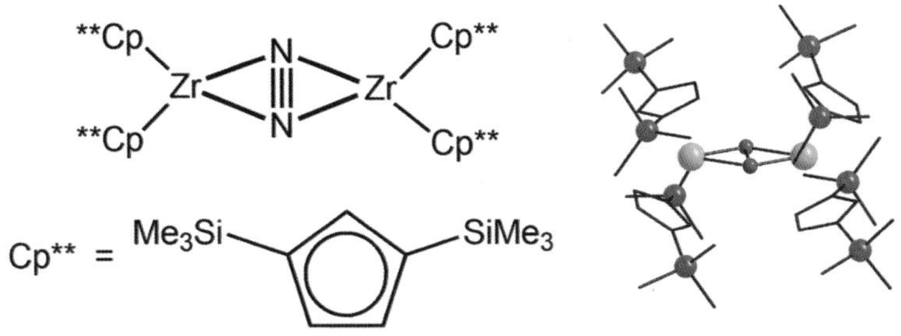

Figure 3.4   The chemical diagram and molecular structure of (**65**)

3.1.2 Trinitrogen

Inverse coordination complexes with trinitrogen units as the coordination centre are illustrated by the azido complex [(μ_2-N$_3$) Cu$_2$(1,10-phenanthroline)$_2$(acetylacetonato)$_2$]$^+$ (**66**; Figure 3.5) [116]; several other examples are known. The azido molecule is linear and links two copper atoms in a transoid arrangement.

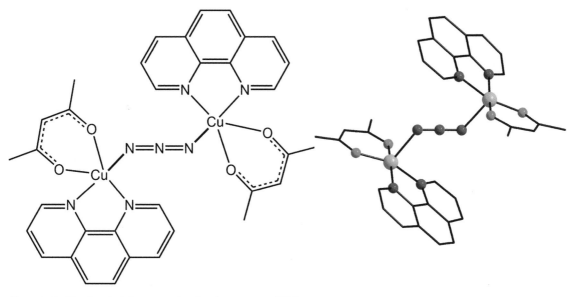

Figure 3.5 The chemical diagram and molecular structure of (**66**)

3.2 Naked phosphorus as the inverse coordination centre

3.2.1 Diphosphorus groups

A rather complicated class of inverse coordination complexes derived from tetrahedrane molecules made of a diphosphorus unit associated with a dimetallic group exists, as illustrated by the compound $[(\mu_2\text{-}P_2)\{\eta^2\text{-}Cr(Cp)(CO)_2\}_2]$ (**67**; Figure 3.6) [117]. While this is not classified as an inverse coordination complex, the donor properties of the phosphorus atoms allow for the decoration of this molecule with metal-containing moieties resulting in the formation of binuclear inverse coordination complexes, as illustrated by the compound $[(\mu_4\text{-}P_2)\{\eta^2\text{-}Cr(Cp)(CO)_2\}_2\{\eta^1\text{-}Cr(CO)_5\}_2]$ (**68**; Figure 3.7) [118]. There are two

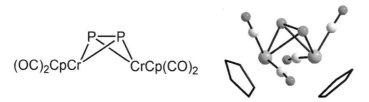

Figure 3.6 The chemical diagram and molecular structure of (**67**)

Figure 3.7 The chemical diagram and molecular structure of (**68**)

distinct modes of attachment for the chromium atoms, which are via a single Cr–P interaction or two such interactions.

A different type of diphosphorus-centred inverse coordination complex is illustrated by the main group element compound $[(\mu_4\text{-}P_2)Bi_4(\mu_2\text{-}Cl)_4\{CH(SiMe_3)_2\}_4]$ (**69**; Figure 3.8) [58], whereby the two pairs of bismuth atoms are linked via a P_2 unit, forming four Bi–P bonds.

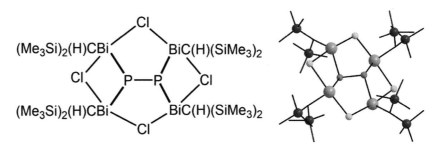

Figure 3.8 The chemical diagram and molecular structure of (**69**)

3.2.2 Triphosphorus groups

Cyclic-P_3 rings form bimetallic triple-decker sandwich complexes, especially when the metal sites are protected by bulky ligands, such as 1,1,1-tris(2-diphenylphosphinoethyl)amine (triphos) [119], and organic π-ligands, such as cyclopentadienyls, as in the compound $[(\mu\text{-}cyclo\text{-}P_3)\{Ni(\eta^5\text{-}C_5Bu^t_3)\}_2]^-$ (**70**; Figure 3.9) [120]. In Figure 3.9, the two nickel centres lie to either side of the P_3 ring, forming interactions with each of the phosphorus atoms.

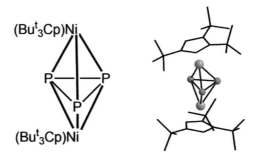

Figure 3.9 The chemical diagram and molecular structure of (**70**)

Trigonal-pyramidal complexes with a *cyclo*-P$_3$ base are also known, such as in the heterobimetallic structure of $[(\mu_4\text{-}cyclo\text{-}P_3)\{\eta^3\text{-}W(Cp^*)(CO)_2\}\{\eta^1\text{-}Mn(Cp)(CO)_2\}_3]$ (**71**; Figure 3.10) [121]. In this example, the tungsten centre sits atop the P$_3$ ring, forming interactions with each of the phosphorus atoms.

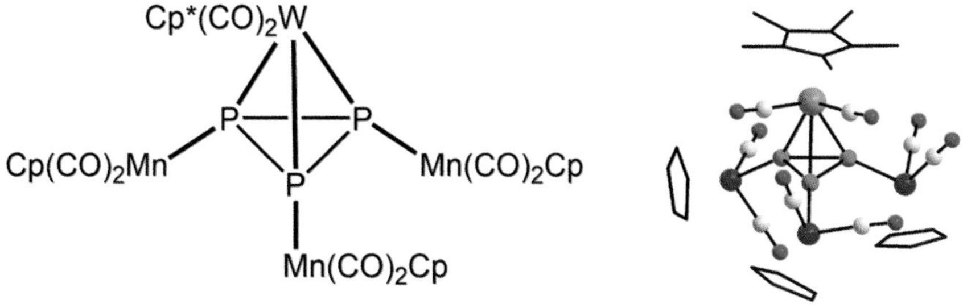

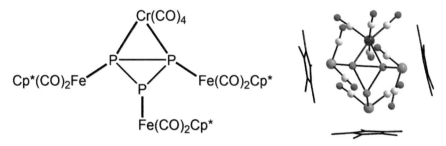

Figure 3.10   The chemical diagram and molecular structure of (**71**)

A distinct topology is observed in the heterobimetallic complex $[(\mu_4\text{-}cyclo\text{-}P_3)\{\eta^2\text{-}Cr(CO)_4\}\{\eta^1\text{-}Fe(Cp^*)(CO)_2\}_3]$ (**72**; Figure 3.11) [122]. In Figure 3.11, the three iron centres can be thought to occupy the base of a tetrahedron with the chromium at the apex, but forming two Cr–P interactions only, in contrast to that seen in Figure 3.10.

Figure 3.11   The chemical diagram and molecular structure of (**72**)

3.2.3 Tetraphosphorus groups

White phosphorus comprises tetrahedral P_4 molecules. Tetrahedral P_4 reacts with (organo)metallic compounds and can remain intact while coordinating to metal centres as a ligand. P_4 can function as a bridge in bimetallic inverse coordination complexes or may undergo a metal-promoted "activation", with cleavage of one, two, or three of the initial P–P bonds leading to the stabilisation of partially opened, butterfly, planar-ring, or chain-like tetraphosphorus units in metal complexes [123].

Intact P_4 molecules usually occur as coordination centres in bimetallic inverse coordination complexes, such as in the $[(\mu_2\text{-}P_4)\{\eta^1\text{-}Ru(Cp)(PPh_3)_2\}_2]^{2+}$ dication (73; Figure 3.12) [124] and in the copper complex $[(\mu_2\text{-}cyclo\text{-}P_3)\{\eta^2\text{-}Cu\{N(C_6H_3Pr^i_2\text{-}2,6)C(Me)CHC(Me)N(C_6H_3Pr^i_2\text{-}2,6)\}_2]$ (74; Figure 3.13) [125]. In compound (73), single connections between the core and each of the ruthenium atoms are found. In the copper compound (74), each copper centre forms two Cu–P bonds so that each phosphorus atom of the P_4 core participates in connecting copper (Figure 3.13).

When one edge of the P_4 tetrahedron is removed, the resulting P_4 moiety can assume different topologies. An example of a partially opened P_4 tetrahedron is a bicyclic species, tetraphosphabicyclo[1.1.0]butane, which can connect two metal atoms via two Cr–P bonds to become an inverse coordination

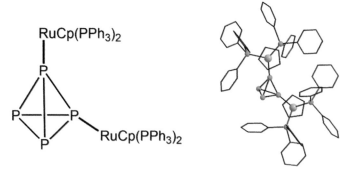

Figure 3.12 The chemical diagram and molecular structure of (**73**)

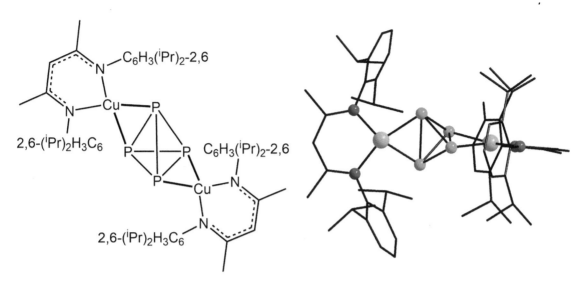

Figure 3.13 The chemical diagram and molecular structure of (**74**)

centre, as illustrated in the organochromium compound $[(\mu_2\text{-}P_4)\{\eta^1\text{-}Cr(Cp^*)(CO)_3\}_2]$ (**75**; Figure 3.14) [126].

Tetraphosphabicyclo[1.1.0]butane can also coordinate to one metal atom to form a chelate complex which becomes an inverse coordination centre when decorated with additional metal-containing moieties. This case is illustrated in the heterobimetallic compound $[(\mu_5\text{-}P_4)\{\eta^2\text{-}Rh(C_6H_3\text{-}(Bu^t)_2\text{-}1,3)CO)\}\{\eta^1\text{-}Cr(CO)_5\}_4]$ (**76**; Figure 3.15) [127], where the rhodium atom is in effect chelated by two of the four available phosphorus atoms.

Figure 3.14 The chemical diagram and molecular structure of (**75**)

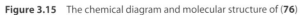

Figure 3.15 The chemical diagram and molecular structure of (**76**)

When two edges are removed from the original P_4 tetrahedron, the resulting tetraphosphorus moiety can adopt various conformations, such as a twisted one leading to butterfly-type complexes, a square-planar conformation leading to a square-pyramid, or a double-decker topology with a P_4 middle deck. Complexation as a P_4-butterfly is illustrated by the organocobalt compound $[(\mu_4\text{-}P_4)\{\eta^1\text{-}Co(Cp^*)(CO)\}_2]$ (77; Figure 3.16) [128]; here, all phosphorus atoms are involved in connecting cobalt centres via two chelating interactions.

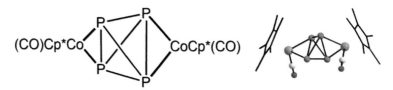

Figure 3.16   The chemical diagram and molecular structure of (77)

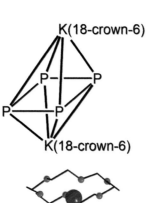

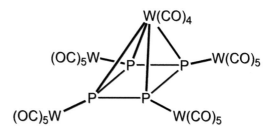

Figure 3.17   The chemical diagram of (78)

A planar P_4 ring that can result from the removal of two P–P edges of the parent P_4 tetrahedron can serve as basis of a square-pyramid with the metal in the apical position, as in the compound $[(\mu_5\text{-}cyclo\text{-}P_4)\{\eta^4\text{-}W(CO)_4\}\{(\eta^1\text{-}W(CO)_5\}_4]$ (78; Figure 3.17) [129].

Planar *cyclo*-tetraphosphorus groups can also form double-decker inverse coordination complexes, with the metal atoms on various sides of the ring. Two examples can be cited: the alkali metal complexes $[(\mu_2\text{-}cyclo\text{-}P_4)\{\eta^4\text{-}ML(18\text{-crown-6})\}_2]$, where M = K (79; Figure 3.18) and Rb [130] in which the metal ions sit to either side of the central ring, and $[(\mu_2\text{-}cyclo\text{-}P_4)\{\eta^2\text{-}U(\eta^5\text{-}Cp^*)$

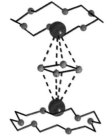

Figure 3.18   The chemical diagram and molecular structure of (79)

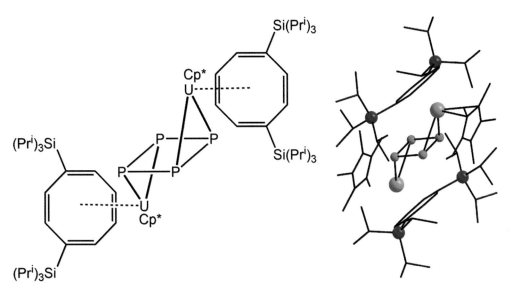

Figure 3.19   The chemical diagram and molecular structure of (**80**)

(η^8-$C_8H_6(SiMe_3)_3$)}$_2$] (**80**; Figure 3.19) [131], where the metal ions are situated over two of the edges, again to either side.

A very rare complex constructed about a planar P_4 ring is the hexanuclear iron compound [(μ_6-*cyclo*-P_4)$Fe_6(CO)_{18}$] (**81**; Figure 3.20) [132], in which each edge of the P_4 ring is alternatively bridged by one or two iron atoms, with the $Fe_2(CO)_6$ residues lying to either side of the P_4 ring.

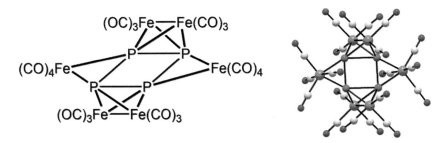

Figure 3.20   The chemical diagram and molecular structure of (**81**)

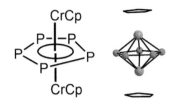

Figure 3.21 The chemical diagram and molecular structure of (**82**)

3.2.4 Pentaphosphorus groups

As a result of the isolobal analogy [133] between the $\equiv P^{III}$ and $\equiv CH$ groups, phosphorus forms five-membered ring complexes similar to the metallocenes. Triple-decker compounds with planar *cyclo*-P_5 centres, LM(*cyclo*-P_5)ML, are well known with a series of metals, and those with M = Cr (**82**; Figure 3.21) [134] present a splendid case of inverse coordination with each chromium atom interacting with each of the five phosphorus atoms, above and below the P_5 plane, akin to that seen in compound (**79**) (Figure 3.18).

A related structure is pentaphosphaferrocene, $(CpFe)(\eta^5$-$P_5)$, a half-sandwich analogue of compound (**82**). In this species, the phosphorus atoms retain electron-donor properties and can act as ligands to form complexes in which the cyclopentaphosphane ring becomes an inverse coordination centre. An example is found in $[(\mu_3$-*cyclo*-$P_5)\{\eta^5$-Fe(Cp*)\}\{\eta^1$-Cr(CO)$_5\}_2]$ (**83**; Figure 3.22) [135], where the iron atoms form bonding interactions with each of the five atoms of the P_5-ring, represented as a single line.

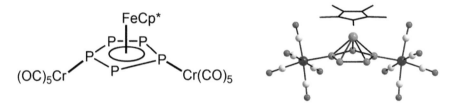

Figure 3.22 The chemical diagram and molecular structure of (**83**)

3.2.5 Hexaphosphorus groups

Cyclo-P_6-based triple-decker complexes have been described for a number of elements and the structures feature planar P_6 rings with D_{6h} symmetry as the inverse coordination centres. This is illustrated by $[(\mu_2$-$P_6)\{\eta^6$-Mo(Cp)\}_2]$ (**84**; Figure 3.23) [136], where the molybdenum atom interacts with each of the six phosphorus atoms, resembling the examples highlighted in inverse coordination complexes (**71**) (Figure 3.10), (**79**) (Figure 3.18), and (**82**) (Figure 3.21), constructed about smaller rings; this is represented by a single line in the chemical diagram of Figure 3.23.

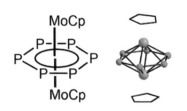

Figure 3.23 The chemical diagram and molecular structure of (**84**)

A notable variation of the aforementioned is seen in the titanium inverse coordination complex, [(μ₂-*cyclo*-P₆){η³-Ti(Cp*)}₂] (**85**; Figure 3.24), in which the central *cyclo*-P₆ moiety adopts a unique non-planar, chair-shaped P₆ ring [137] and in which the titanium atoms connect to alternating phosphorus atoms in the ring. Another variant of the P₆ unit is a bicyclic P₆ centre as found in the structure of [(μ₂-P₆){η³-Th(η⁵-C₅H₃Buᵗ₂)₂}] (**86**; Figure 3.25) [138], with four of the phosphorus atoms linked to thorium.

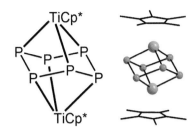

Figure 3.24 The chemical diagram and molecular structure of (**85**)

Figure 3.25 The chemical diagram and molecular structure of (**86**)

3.2.6 Heptaphosphorus groups

A phosphanortricyclane [P₇]³⁻ anion acts as an inverse coordination centre to form some organo-germanium, -tin, and -lead compounds, [(μ-P₇)(η¹-MLₓ)₃], as exemplified by M = Pb and L = Me (**87**; Figure 3.26) [139] with single points of contact between the core and three lead atoms.

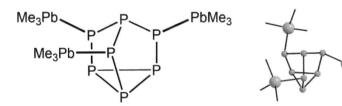

Figure 3.26 The chemical diagram and molecular structure of (**87**)

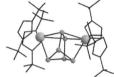

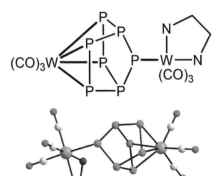

Figure 3.27 The chemical diagram and molecular structure of (**88**)

A phosphanorbornadiene is also known to form inverse coordination complexes with a P₇ centre, for example in the tungsten trianion, [(μ-P₇)(η⁴-W(CO)₃)(η¹-W(CO)₃(H₂NCH₂CH₂NH₂))]³⁻ (**88**; Figure 3.27) [140], where the core links two disparate tungsten centres in two distinct manners, which are via one or four W–P interactions.

3.2.7 Octaphosphorus groups

Octaphosphacuneane forms an organonickel inverse coordination complex, $[(\mu_2\text{-}P_8)(\eta^2\text{-}Ni\{C_5H_2(Bu^t)_3\}_2]^{2-}$ (89; Figure 3.28) [121], with each nickel centre chelated by the core. Several tetranuclear complexes with P_8 realgar type coordination centres have been reported, including a compound of samarium, that is $[(\mu_4\text{-}P_8)\{\eta^2\text{-}Sm(Cp^*)_2\}_4]$ (90; Figure 3.29) [141], whereby four of the phosphorus atoms serve as bridges between samarium atoms arranged in a square.

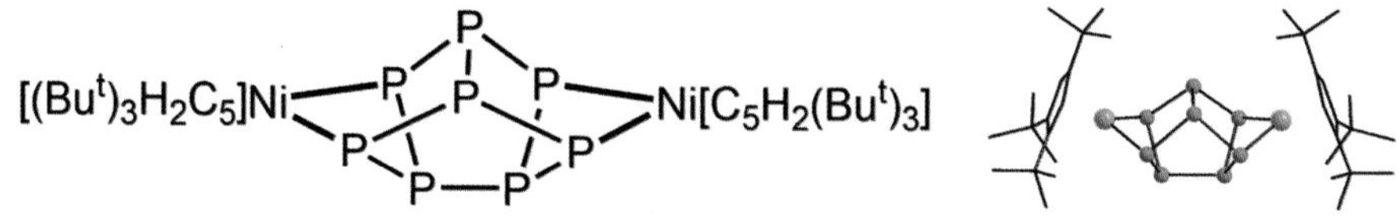

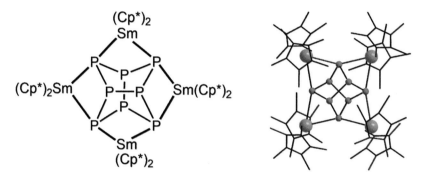

Figure 3.28   The chemical diagram and molecular structure of (**89**)

Figure 3.29   The chemical diagram and molecular structure of (**90**)

An octaphosphorus molecule with a rare topology functioning as an inverse coordination centre is now highlighted. The topology is based on a planar four-membered P_4 ring with four pendant PBu^t_2 groups, and serves to connect two $Ni(CO)_2$ residues in a chelating mode in the structure of $[(\mu_2\text{-}cyclo\text{-}P_4(PBu^t_2)_4)\{\eta^2\text{-}Ni(CO)_2\}_2]$ (**91**; Figure 3.30) [142].

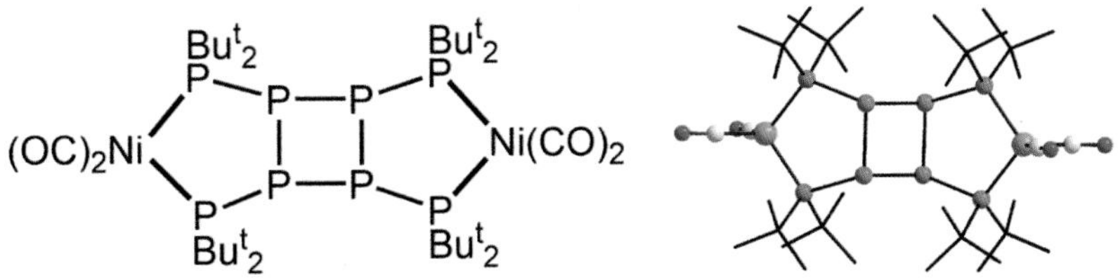

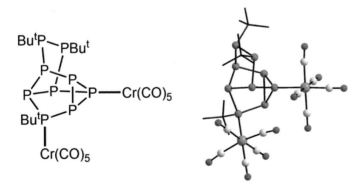

Figure 3.30   The chemical diagram and molecular structure of (**91**)

3.2.8 Nonaphosphorus and larger P$_n$ groups

It has been noted that sometimes reactions of tetrahedral P$_4$ molecules with various metallic reagents capable of cleaving P–P bonds produce fragments which then self-aggregate into larger groups containing 9, 10, 11, 12, 14, 16, and even 24 phosphorus atoms, forming spectacular inverse coordination complexes [143,144]. For illustrative purposes, a small selection is presented here.

Inverse coordination complexes with P$_9$ coordination centres are rare but can be illustrated with the chromium compound $[(\mu\text{-}P_9Bu^t_3)\{\eta^1\text{-}Cr(CO)_5\}_2]$ (**92**; Figure 3.31) [145] where one Cr(CO)$_5$ group is connected to a naked phosphorus donor and the other to a phosphane-phosphorus atom.

Figure 3.31   The chemical diagram and molecular structure of (**92**)

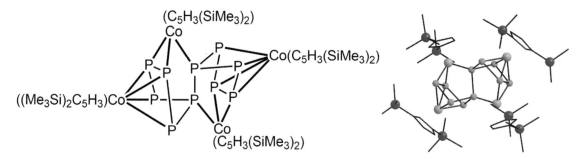

Figure 3.32 The chemical diagram and molecular structure of (**93**)

Complexes with P_{10} coordination centres are more common than those with P_9 centres. A good example is found in the cobalt compound $[(\mu_4\text{-}P_{10})(\eta^3\text{-}CoCp^{Si})_2(\eta^4\text{-}CoCp^{Si})_2]$ (**93**; Figure 3.32) [146], where Cp^{Si} is $C_5H_3(SiMe_3)_2\text{-}1,3$, whereby each cobalt centre connects to the inverse coordination centre via three or four Co–P interactions.

A second example with a P_{10} core is seen in the heterobimetallic iron-samarium complex $[(\mu_4\text{-}P_{10})\{\eta^2\text{-}\{Sm(Cp^*)\}\}_2\{\eta^4\text{-}Fe(Cp^*)\}_2]$ (**94**; Figure 3.33) [147], where each iron centre is connected to the core by four Fe–P bonds.

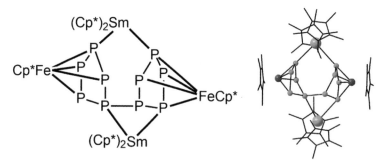

Figure 3.33 The chemical diagram and molecular structure of (**94**)

4 ORGANIC MOLECULES AS INVERSE COORDINATION CENTRES

4.1 Nitrogen heterocycles as inverse coordination centres

Organic nitrogen heterocycles play an extraordinary role as ligands in traditional coordination chemistry, owing to the potent donor properties of the nitrogen atoms for metal centres. They can also be considered to function as inverse coordination centres if the heterocyclic ring contains at least two heteroatoms required for bridging metal centres decorating the ring. Alternatively, the cooperation of heteroatoms in a molecule (e.g., pyridine) even with only one donor for metals with some functional groups (e.g., pyridyl, alkylamino, or carboxylato moieties) allows the resulting species to act as an inverse coordination centre. The number and variety of such inverse coordination complexes is bewildering, and only a small selection is presented here to illustrate the enormous range of possibilities.

4.1.1 Six-membered heterocycles

Being compact and having two nitrogen heteroatoms in 1,4-positions, one of the most efficient nitrogen heterocycles functioning as an inverse coordination centre is pyrazine, which is able to connect two metal atoms as shown in the binuclear molybdenum(VI) complex $(\mu_2$-pyrazine)$[Mo(=O)Cl_4]_2$ (**95**; Figure 4.1) [148].

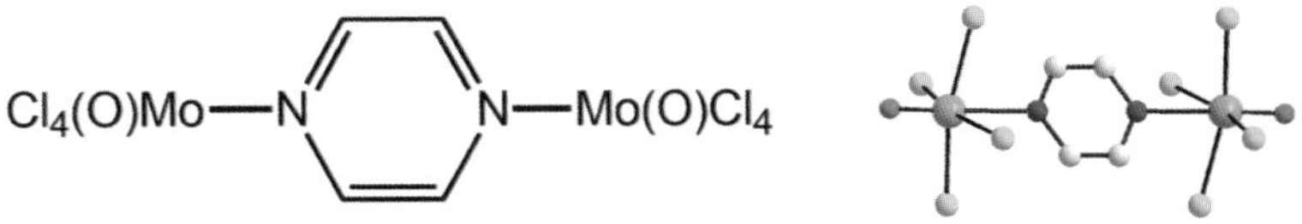

Figure 4.1 The chemical diagram and molecular structure of (**95**)

As mentioned in the preamble, functionalisation of the nitrogen heterocycles is possible. In the case of pyrazine, two 2-pyridyl substituents can be appended in different positions leading to two types of inverse coordination centres: one with a *syn*-configuration, for example in the rhenium(I) complex (μ_2-1,2-dipyridylpyrazine)[Re(CO)$_3$Cl]$_2$ (**96**; Figure 4.2) [149], and another with an *anti*-configuration, as found in the tetracationic binuclear nickel(II) complex {(μ_2-1,4-dipyridylpyrazine)[Ni(OH$_2$)$_4$]$_2$}$^{4+}$ (**97**; Figure 4.3) [150]. In each molecule, the metal atom is chelated by pyrazine- and pyridyl-nitrogen donor atoms to form five-membered chelate rings.

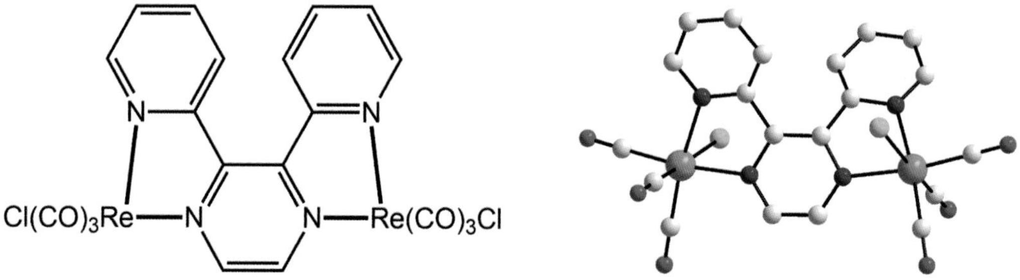

Figure 4.2   The chemical diagram and molecular structure of (**96**)

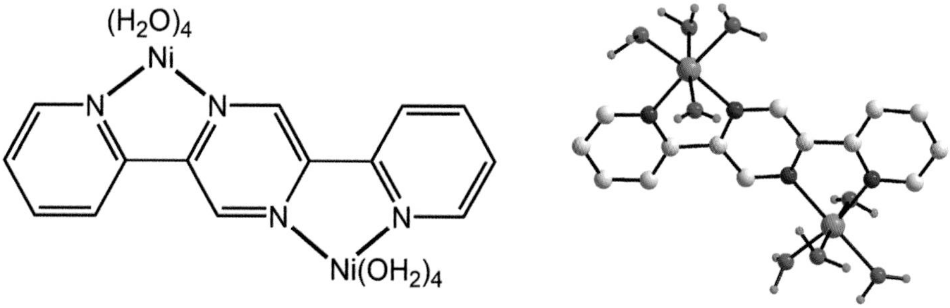

Figure 4.3   The chemical diagram and molecular structure of (**97**)

Four pyridyl substituents attached to the pyrazine heterocycle afford the formation of aesthetically pleasing inverse coordination complexes, as exemplified by dicationic {(μ_2-2,3,5,6-tetrapyridylpyrazine)[PtCl]$_2$}$^{2+}$ (**98**; Figure 4.4) [151], where the inverse coordination centre functions as a hexadentate ligand, forming four five-membered chelate rings leading to square-planar geometries for the platinum(II) centres. A similar inverse coordination centre and mode of association, but with aminomethyl groups rather than 2-pyridyl substituents, are possible as seen in the structure of the trigonal-bipyramidal zinc(II) species (μ_2-2,3,5,6-methylaminopyrazine)[ZnCl$_2$]$_2$ (**99**; Figure 4.5) [152].

Pyrimidine is another versatile heterocyclic molecule capable of acting as an inverse coordination centre. In its unsubstituted "naked" form, it generates examples such as the dianionic octahedral rhenium(IV) complex, {(μ_2-pyrimidine)[ReBr$_5$]$_2$}$^{2-}$ (**100**; Figure 4.6) [153], and the dianionic square-planar platinum(II) complex, {(μ_2-pyrimidine)[PtCl$_3$]$_2$}$^{2-}$ (**101**; Figure 4.7) [154].

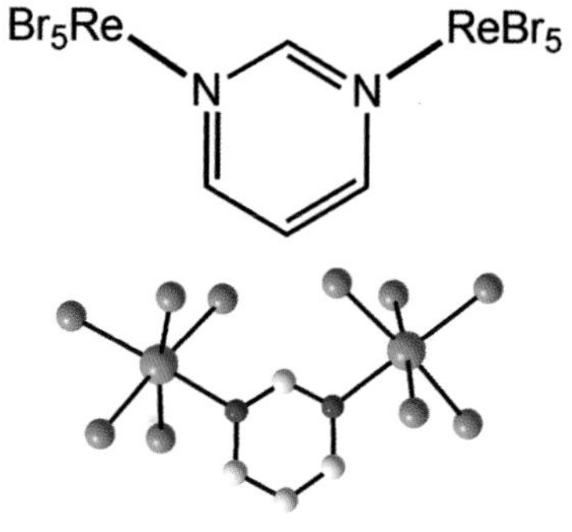

Figure 4.6 The chemical diagram and molecular structure of (**100**)

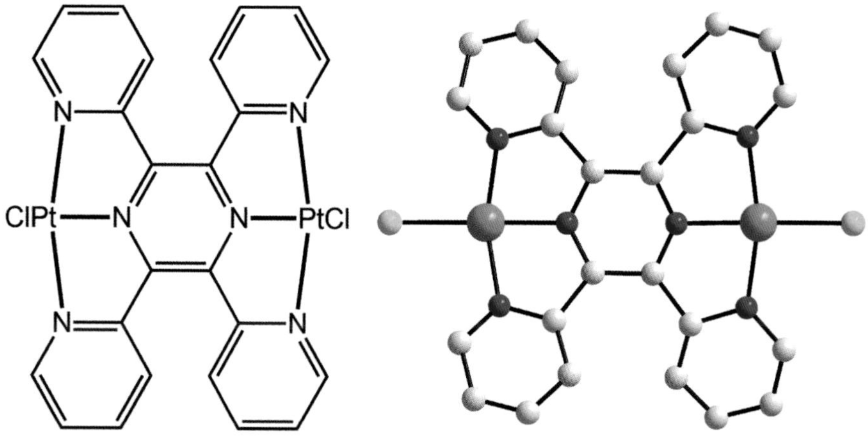

Figure 4.4 The chemical diagram and molecular structure of (**98**)

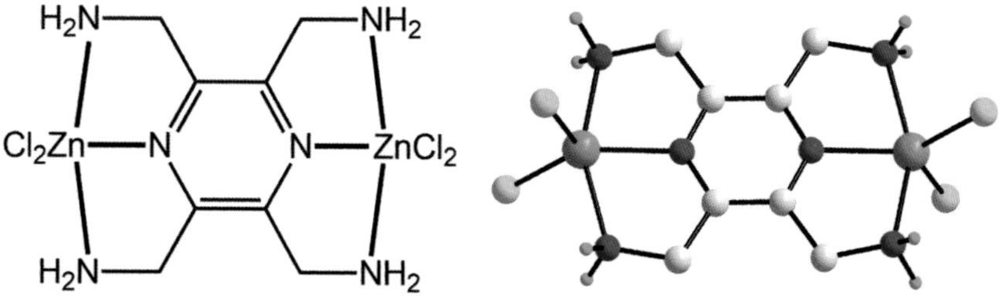

Figure 4.5 The chemical diagram and molecular structure of (**99**)

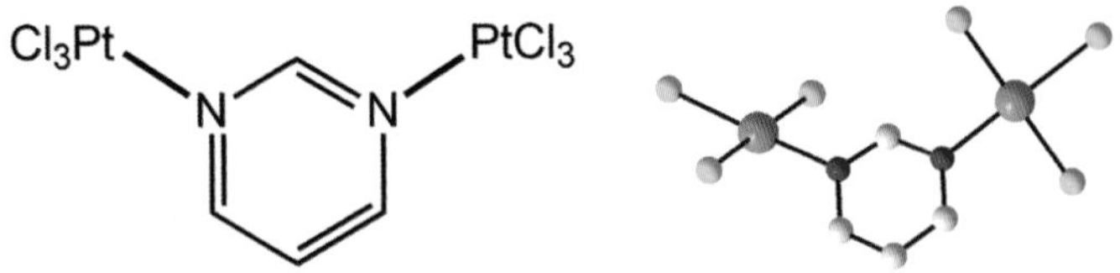

Figure 4.7   The chemical diagram and molecular structure of (**101**)

Functionalisation with two carboxylato substituents leads to inverse coordination complexes like the dicationic platinum(II) complex {(μ₂-4,6-dicarboxylato-pyrimidine)[Pt(NH₃)₂]₂}²⁺ (**102**; Figure 4.8) [155].

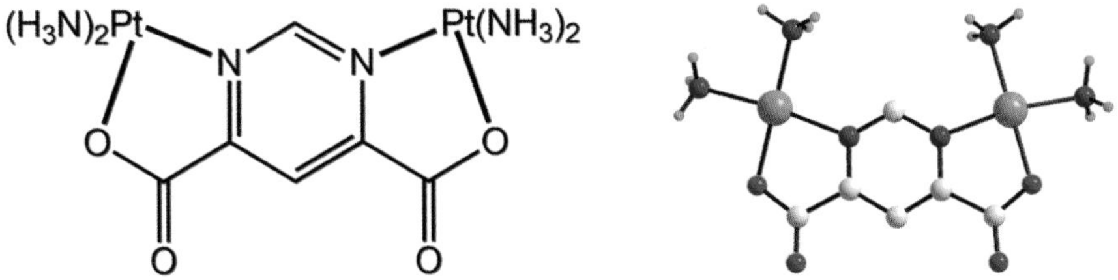

Figure 4.8   The chemical diagram and molecular structure of (**102**)

The isomer of azine, 1,2-diazine (pyridazine), has rarely been exploited as an inverse coordination centre, but a complex illustrating this ability is found in the dicationic silver compound {(μ₂-pyridazine)[Ag(2,2′:6′,2″-terpyridine)]₂}²⁺ (**103**; Figure 4.9) [156].

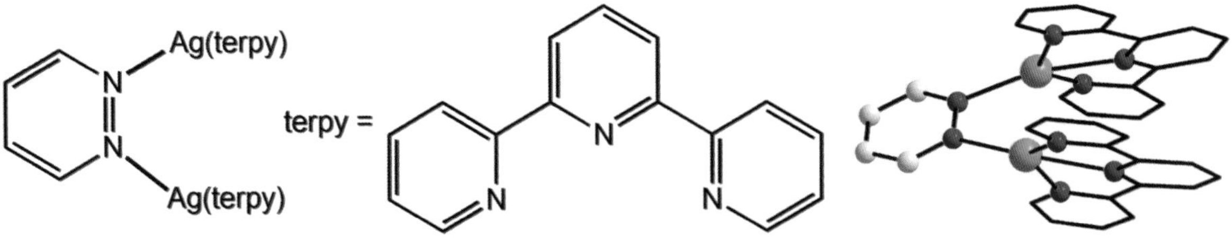

Figure 4.9   The chemical diagram and molecular structure of (**103**)

The 1,3,5-triazine heterocycle can employ only two, or all three, of the potential nitrogen donor atoms to attach and bridge metal atoms and thus, functions as a centre for inverse coordination complexes. An example where two nitrogen donors are employed to attach metals is found in binuclear and dianionic {(μ₂-triazine)[ReBr₅]₂}²⁻ (104; Figure 4.10) [153], which may be compared with complex (100) (Figure 4.6). An example where all three nitrogen-donors are employed is seen in trinuclear, (μ₂-triazine)[Pt(PEt₃)Cl₂]₃ (105; Figure 4.11) [157].

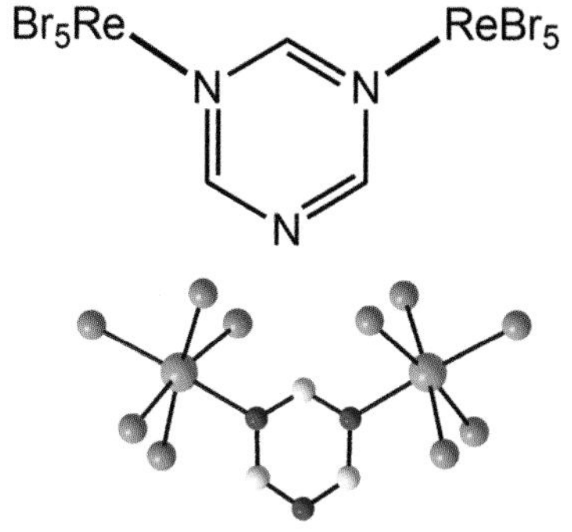

Functional derivatives of triazine, such as those with thiolato substituents, are efficient inverse coordination centres. An example is found in the 2,4,6-trithiolato triazine species whereby all nitrogen donors and the three exocyclic sulphur atoms connect the three metals, as shown in the trinuclear rhodium(III) species (μ₃-2,4,6-trithiolatotriazine)[Rh(C₅Me₅)Cl]₃ (106; Figure 4.12) [158].

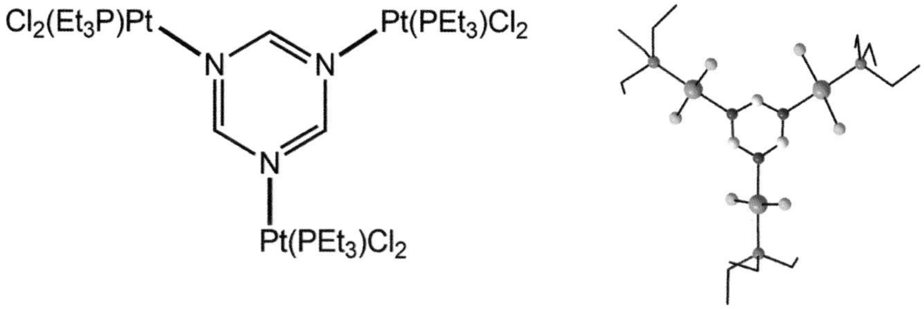

Figure 4.10 The chemical diagram and molecular structure of (**104**)

Figure 4.11   The chemical diagram and molecular structure of (**105**)

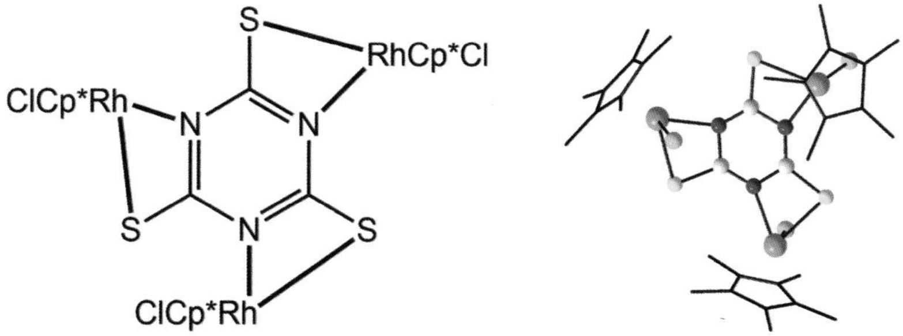

Figure 4.12   The chemical diagram and molecular structure of (**106**)

As for pyridazine, 1,2,4,5-tetrazine (tetrazine) has been rarely used as a coordination centre. An example of a complex centred about 1,2,4,5-tetrazine is illustrated by the binuclear tetracation $\{(\mu_2\text{-tetrazine})[Ru(NH_3)_5]_2\}^{4+}$ (**107**; Figure 4.13) [159], featuring octahedrally-coordinated ruthenium(II) centres occupying 1,4-positions about the ring.

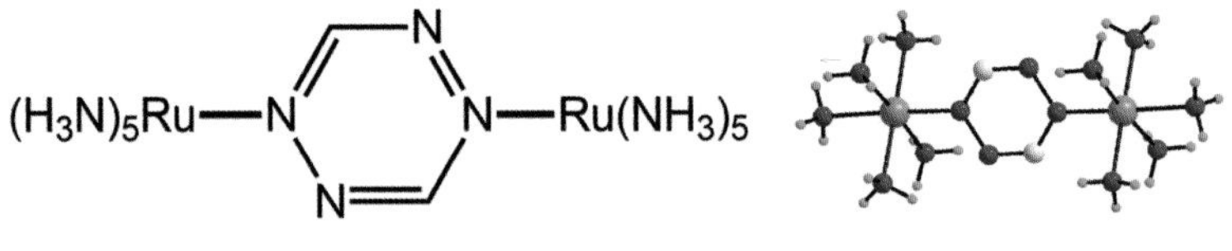

Figure 4.13   The chemical diagram and molecular structure of (**107**)

4.1.2 Bicyclic molecules

The class of compounds where six-membered heterocycles are connected in pairs form a remarkable variety of efficient inverse coordination centres. From a long list of such known compounds, the example given here is the manganese(II) tetracationic complex of 2,2′-bipyrimidine, that is, $\{(\mu_2\text{-2,2′-bipyrimidine})[Mn(OH_2)_5]_2\}^{4+}$ (**108**; Figure 4.14) [160].

Numerous inverse coordination complexes can be derived from 4,4′-bipyridine, which normally tends to form coordination polymers [161]. Two illustrative examples of discrete molecular inverse coordination complexes are found in the transition metal complex of octahedrally-coordinated niobium(V), namely $(\mu_2\text{-4,4′-bipyridine})[Nb(N_3)_5]_2$ (**109**; Figure 4.15) [162], and in the main

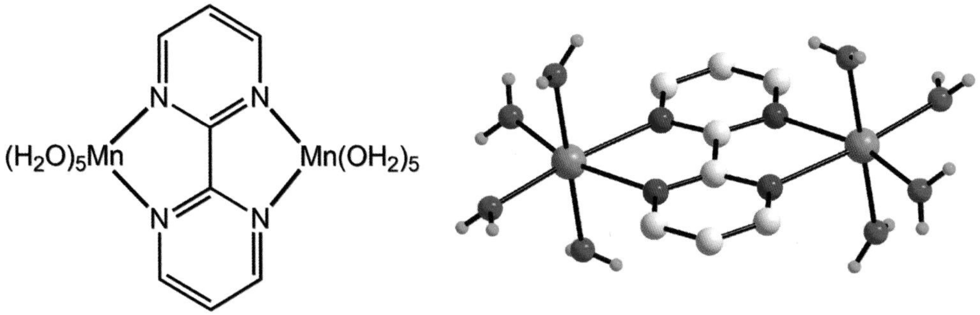

Figure 4.14   The chemical diagram and molecular structure of (**108**)

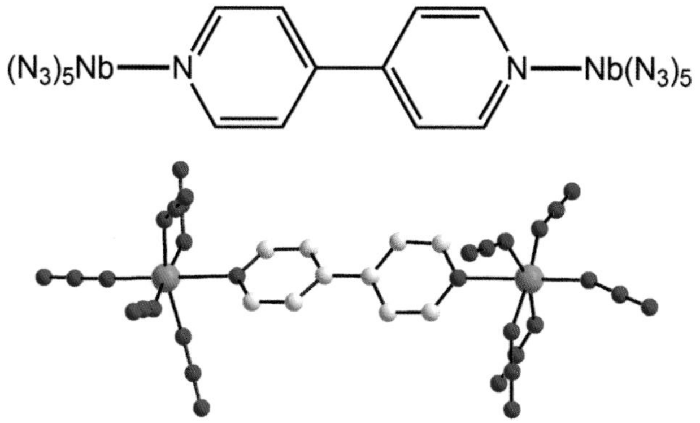

Figure 4.15   The chemical diagram and molecular structure of (**109**)

group element compound of similarly octahedrally-coordinated bismuth(V), that is tetraanionic $\{(\mu_2\text{-}4,4'\text{-bipyridine})[BiCl_5]_2\}^{4-}$ (**110**; Figure 4.16) [163].

Far less common are the inverse coordination complexes derived from isomeric 3,3'-bipyridine. An example is found in the octahedrally-coordinated niobium(V) complex, $(\mu_2\text{-}3,3'\text{-}$ bipyridine)$[Nb(N_3)_5]_2$ (**111**; Figure 4.17) [162], which is an isomer

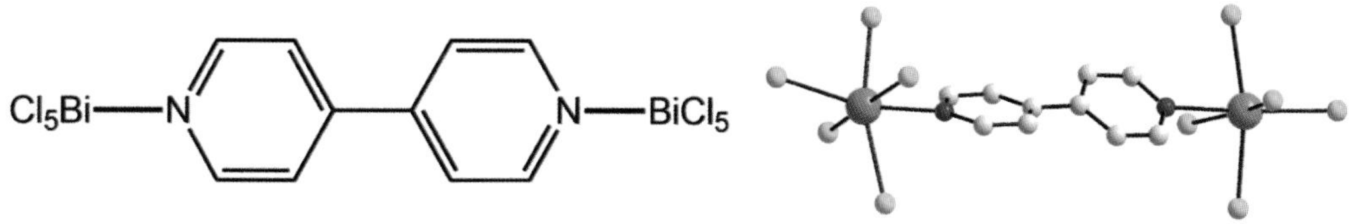

Figure 4.16   The chemical diagram and molecular structure of (**110**)

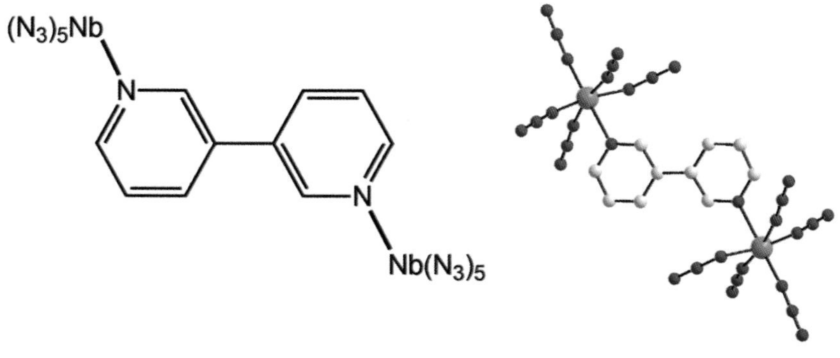

Figure 4.17   The chemical diagram and molecular structure of (**111**)

of compound (**109**) (Figure 4.15) with the common feature that the Nb(N$_3$)$_5$ substituents occupy diagonally opposite positions.

Many inverse coordination complexes are formed with coordination centres in which pyridine moieties are appended to some hydrocarbon skeleton. Examples drawn from this broad class of compound include the organoaluminium compound (μ$_2$-1,2-di-(4-pyridyl)ethane)[AlMe$_3$]$_2$ (**112**; Figure 4.18) [164] and the zinc dithiocarbamato compound (μ$_2$-1,2-di-(4-pyridyl) ethene)[Zn(S$_2$CNEt$_2$)$_2$]$_2$ (**113**; Figure 4.19) [165].

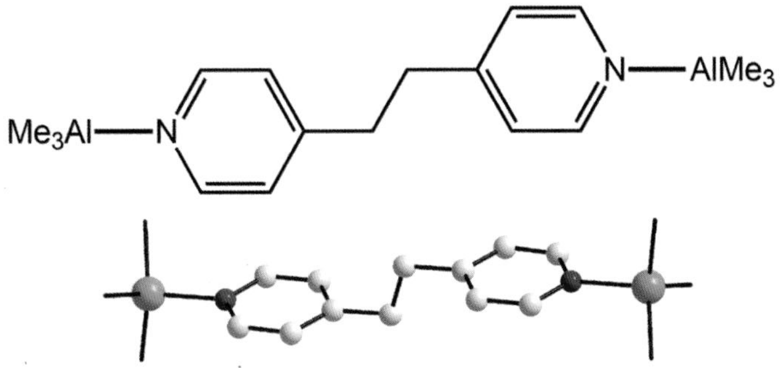

Figure 4.18   The chemical diagram and molecular structure of (**112**)

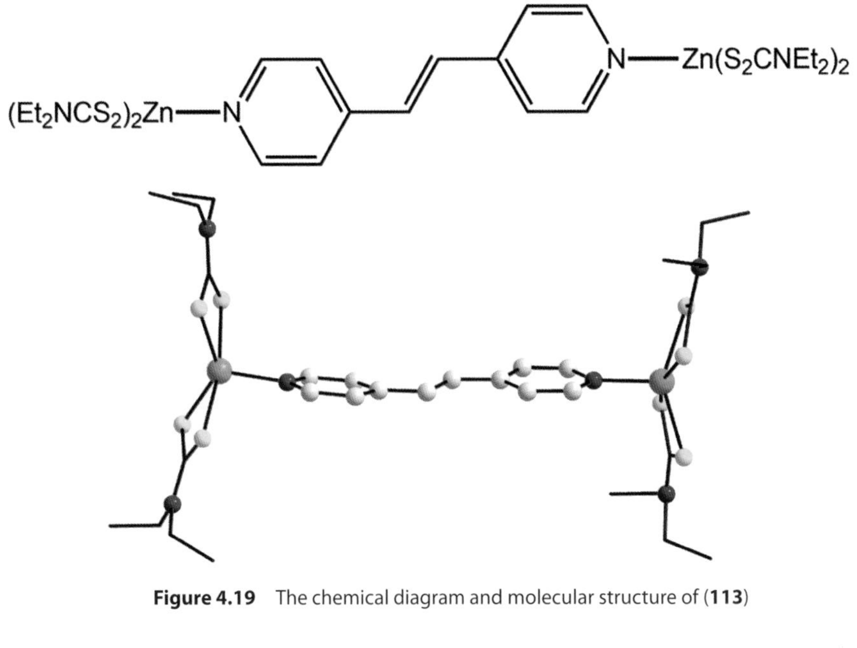

Figure 4.19   The chemical diagram and molecular structure of (**113**)

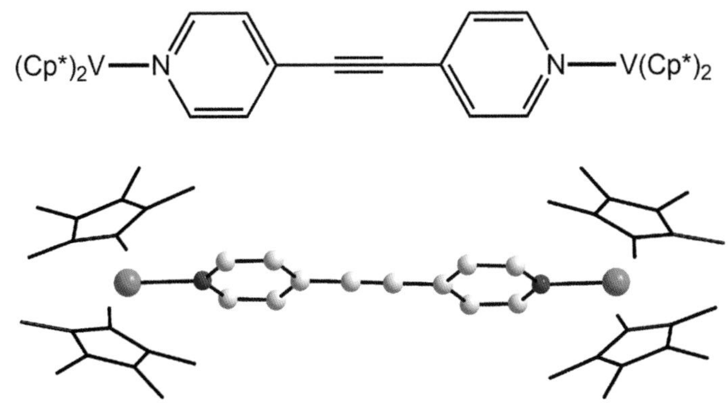

Figure 4.20 The chemical diagram and molecular structure of (**114**)

More exotic linkers are seen in the structures of two organovanadium complexes [166], namely (μ_2-1,2-di-(4-pyridyl) acetylene)[V(Cp*)$_2$]$_2$ (**114**; Figure 4.20) and (μ_2-4,4'-azodipyridine) [V(Cp*)$_2$]$_2$ (**115**; Figure 4.21).

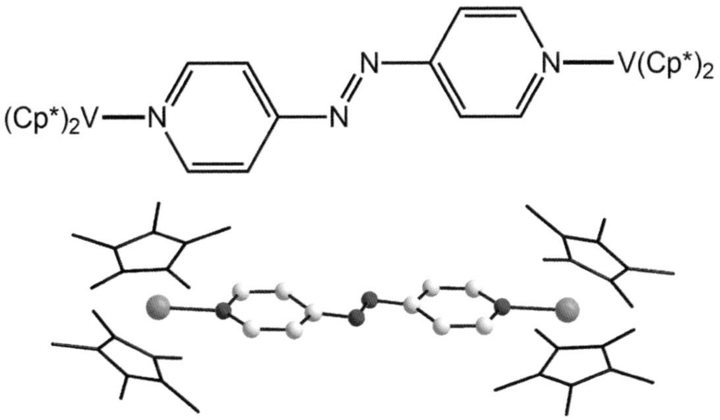

Figure 4.21 The chemical diagram and molecular structure of (**115**)

4.1.3 Miscellaneous

To highlight the wide diversity of centres possible for inverse coordination, several specialised, pyridyl-appended molecules are briefly presented here, suggesting and perhaps inspiring new synthetic possibilities in the field. These chosen examples include binuclear (μ_2-tetrakis(2-pyridyl)cyclobutane)[Cu(NO$_3$)$_2$]$_2$ (**116**; Figure 4.22) [167], with the pyridine moieties attached to a square (cyclobutane) ring, and (μ_2-tetrakis(2-pyridyl)methane) [Mn(1,1,1,5,5,5-hexafluoropentane-2,4-dionato)$_2$]$_2$ (**117**; Figure 4.23) [168], with four pyridine groups attached to the same carbon atom.

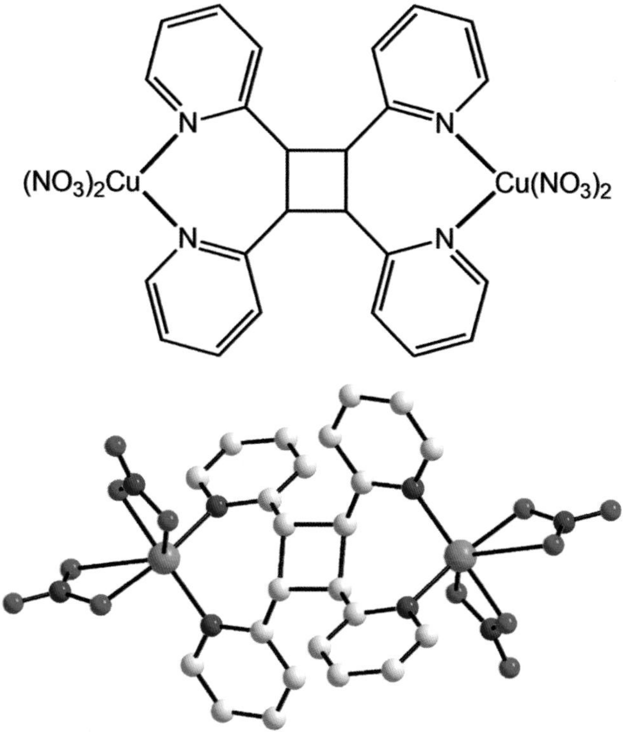

Figure 4.22   The chemical diagram and molecular structure of (**116**)

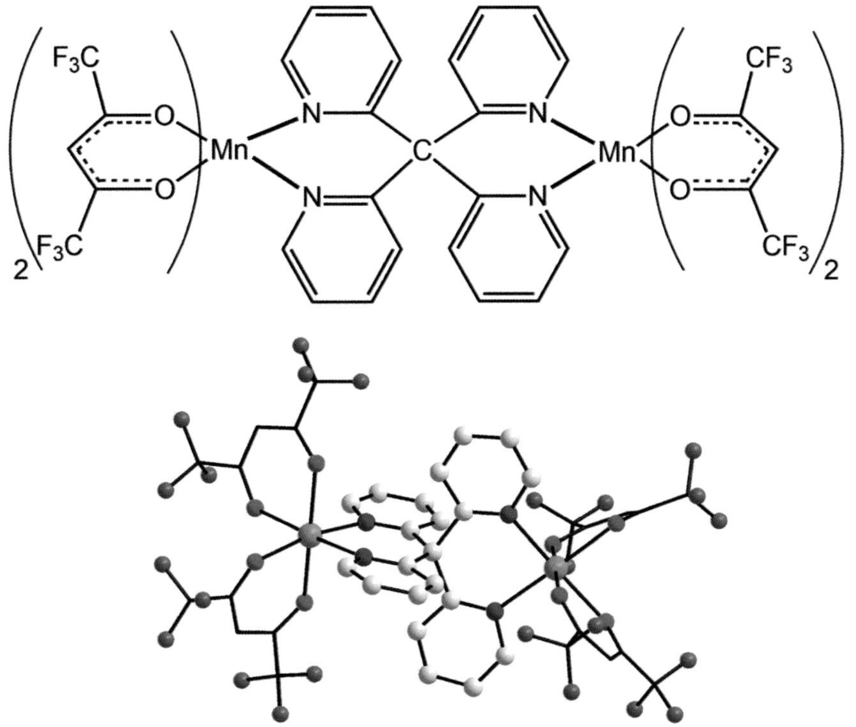

Figure 4.23 The chemical diagram and molecular structure of (**117**)

Larger atoms such as phosphorus and silicon may also function as the central atom as seen in the rather complex molecules of (μ_3-tris(4-pyridyl)phosphane){Zn[(4,5-dichloro-o-phenylenediamine) bis(3-t-butylsalicylidenealdiminaoto)]}$_3$ (**118**; Figure 4.24) [169], with the three pyridyl linkers carried by a phosphorus atom, and in the rhodium(II) paddle-wheel complex, (μ_3-tris(4′-pyridyl) methylsilane)[Rh$_2$(OOCCF$_3$)$_4$(benzene)]$_3$ (**119**; Figure 4.25) [170], with the pyridyl linkers joined at a silicon atom.

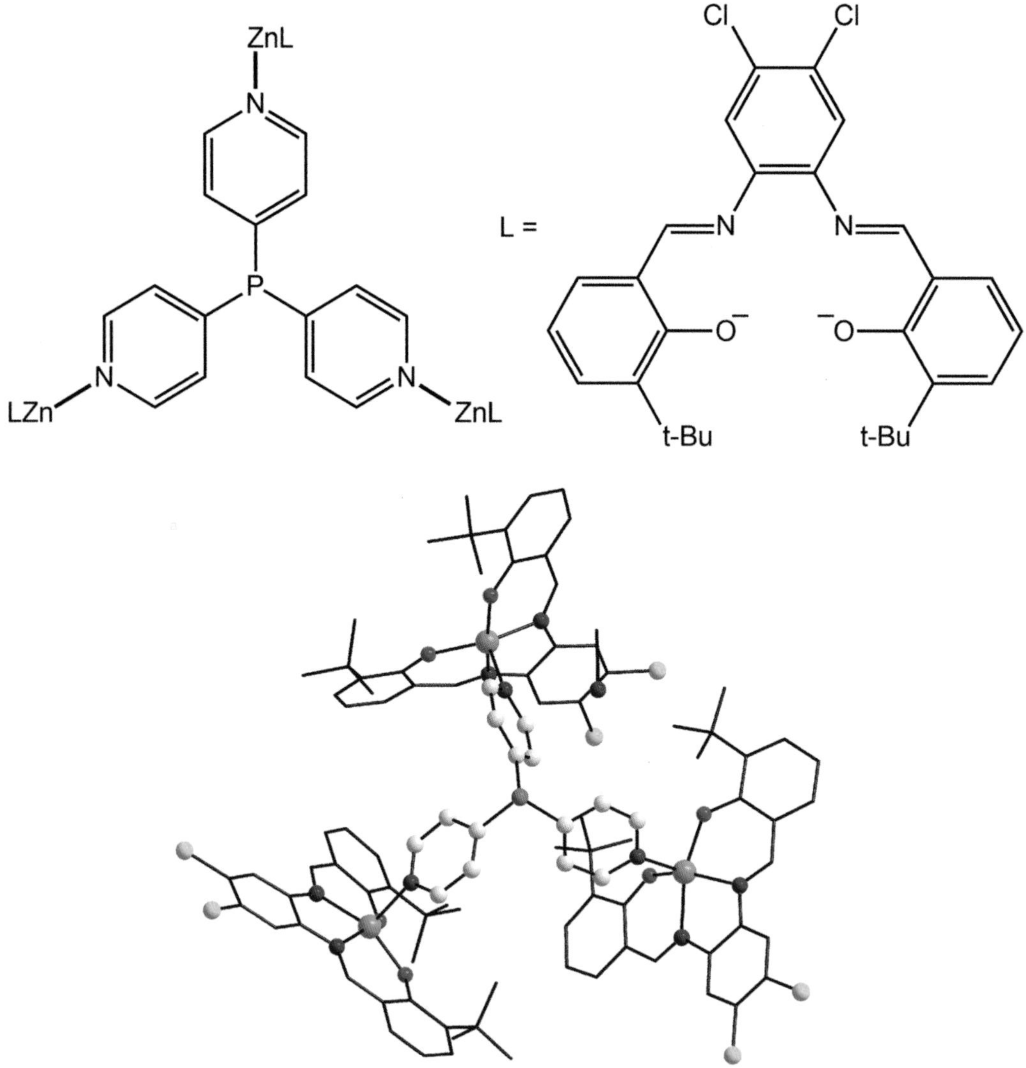

Figure 4.24   The chemical diagram and molecular structure of (**118**)

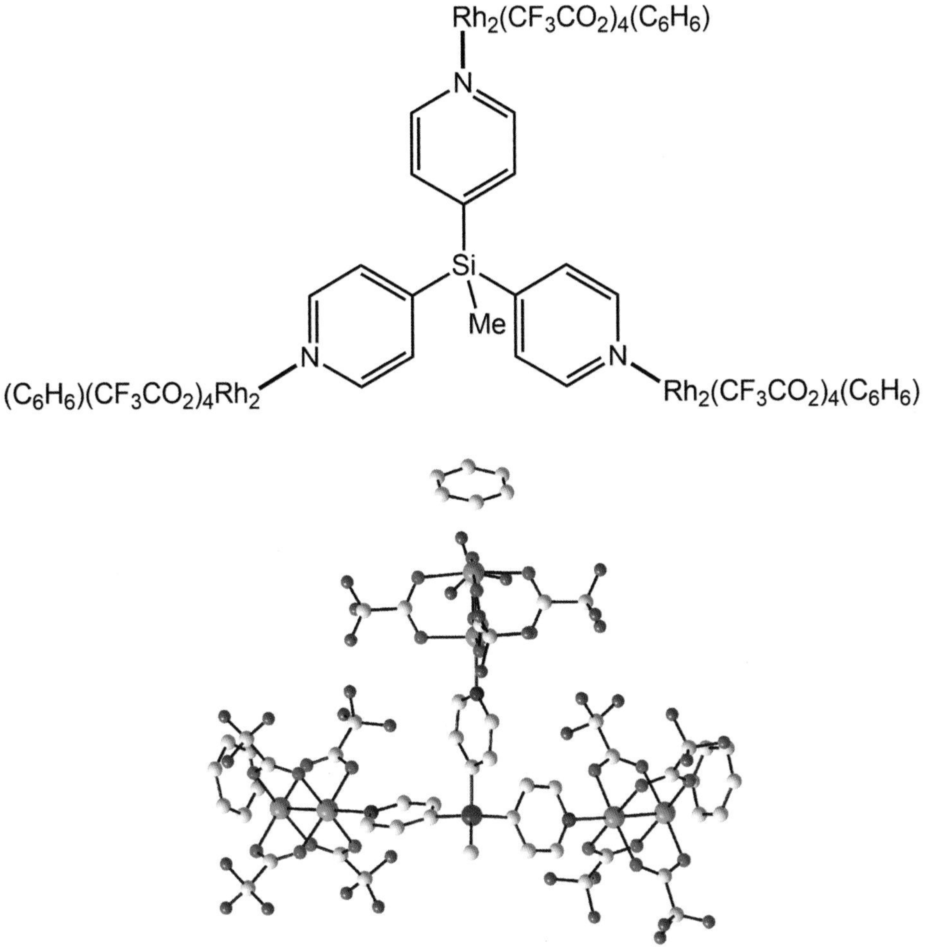

Figure 4.25 The chemical diagram and molecular structure of (**119**)

4.2 Fused-ring systems as inverse coordination centres

Various polynuclear nitrogen heterocycles form inverse coordination complexes. These can be bi-connective triarene derivatives like the organoaluminium compound (μ_2-phenazine) [AlMe$_3$]$_2$ (**120**; Figure 4.26) [171], and that linking the two square-planar platinum(II) entities (μ_2-4,7-phenanthroline)[PtBr$_2$(SMe$_2$)]$_2$ (**121**; Figure 4.27) [172].

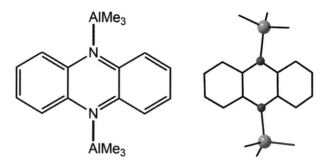

Figure 4.26   The chemical diagram and molecular structure of (**120**)

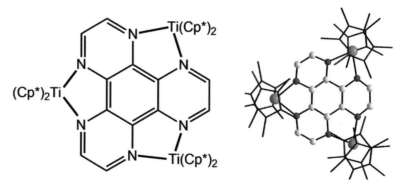

Figure 4.27   The chemical diagram and molecular structure of (**121**)

Tetraarene inversion complexes are capable of connecting three metal centres, such as in the titanocene complex of a derivative of quinoxaline, (μ_2-4a,4b,8a,8b,12a,12b-hexahydrodipyrazino(2,3-f:2′,3′-h)quinoxaline)[TiCp*$_2$]$_3$ (**122**; Figure 4.28) [173].

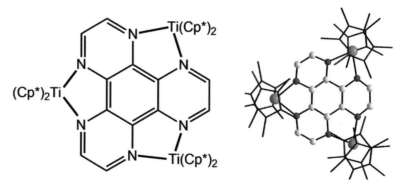

Figure 4.28   The chemical diagram and molecular structure of (**122**)

The quinoxaline inverse coordination centres may be decorated with six additional 2-pyridyl groups available for further coordination; for example, to cadmium as in (μ_3-2,3,6,7,10,11-hexakis(2-pyridyl)dipyrazino[2,3-f:2′,3′-h]quinoxaline) [Cd(NO$_3$)$_2$(OH$_2$)]$_3$ (123; Figure 4.29) [174].

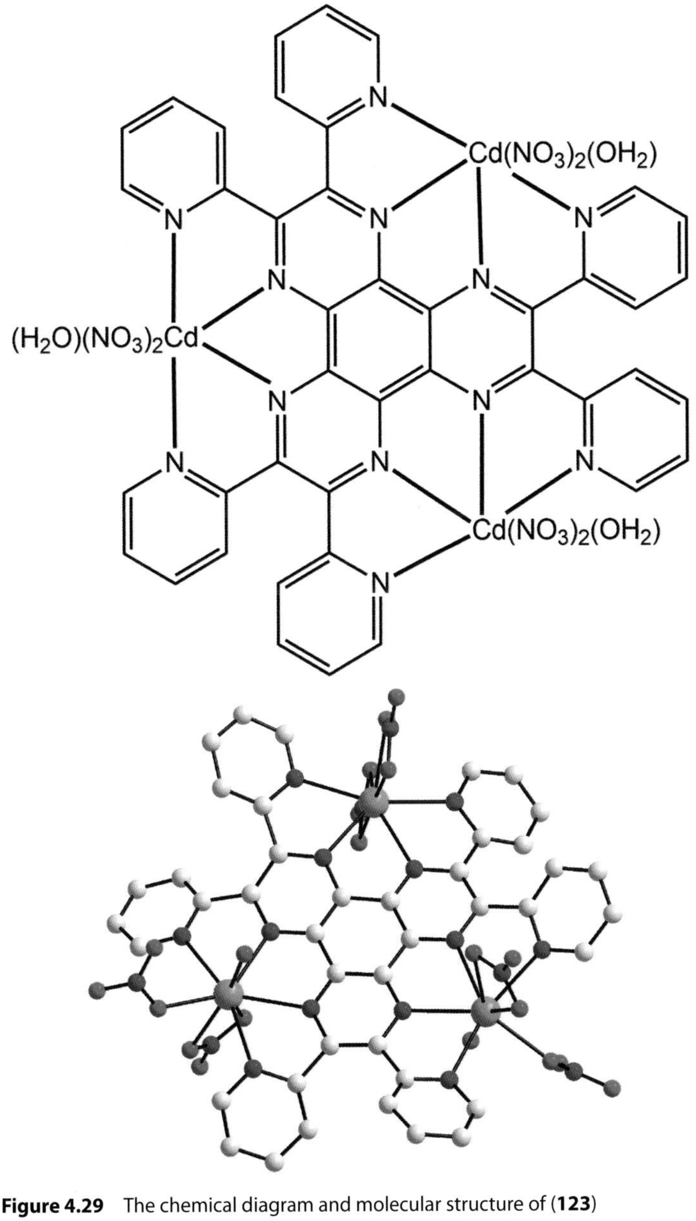

Figure 4.29   The chemical diagram and molecular structure of (123) (note that water–hydrogen atoms are not shown in the molecular diagram)

Pentaarene derivatives can also function as inverse coordination centres, whereby they connect two metal centres, as in the octahedral ruthenium(II) dicationic species, {(μ_2-1,6,7,12-tetra-azaperylene)[Ru(1-isopropyl-4-methylbenzene)Cl]$_2$}$^{2+}$ (124; Figure 4.30) [175].

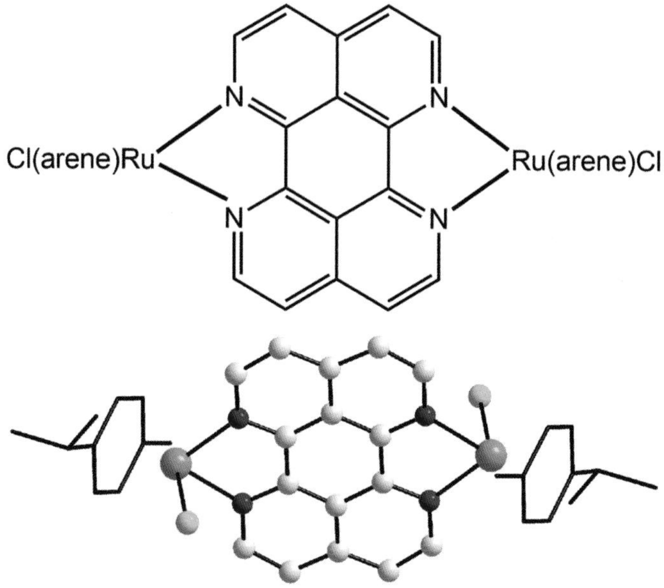

Figure 4.30   The chemical diagram and molecular structure of (**124**)

A spectacular inverse coordination complex constructed about nonaarene is evident in the octahedral ruthenium(II) dicationic species, (μ_2-dibenzoeilatin)[Ru(tetramethyl-2,2′-bipyridyl)$_2$]$_2$ (125; Figure 4.31) [176].

4.3 Saturated six-membered rings as inverse coordination centres

4.3.1 Monocyclic species

The first example in this category features a piperazine core as the inverse coordination centre being connected to two trimethylindium centres in (μ_2-1,4-dibenzylpiperazine)[InMe$_3$]$_2$ (**126**; Figure 4.32) [177].

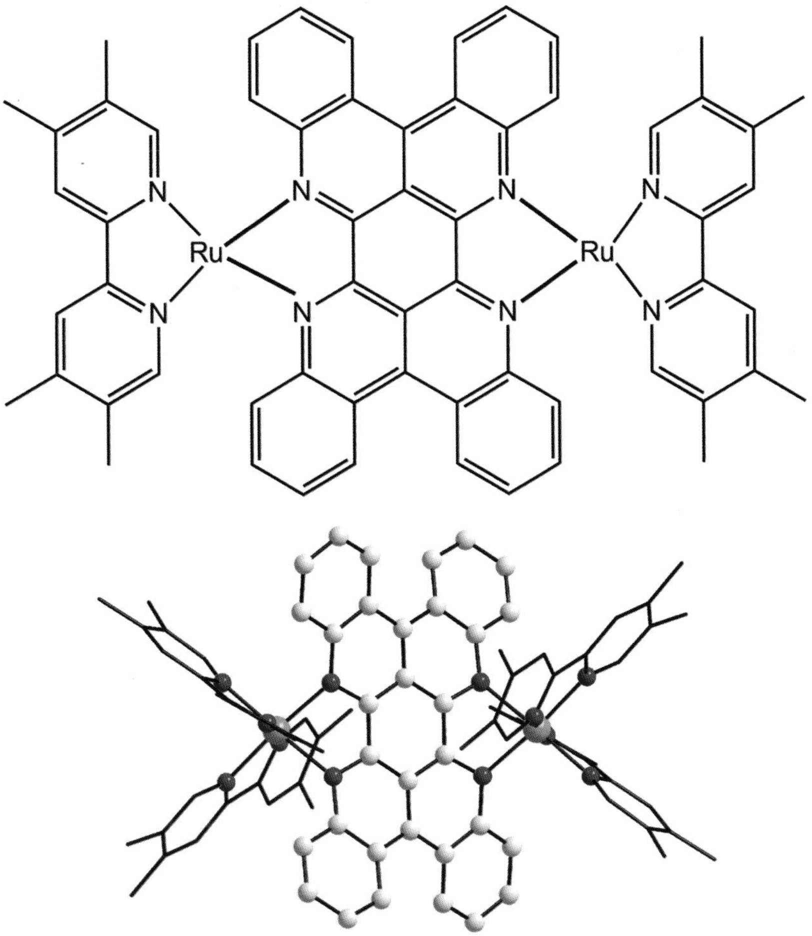

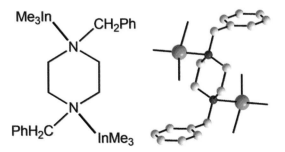

Figure 4.31   The chemical diagram and molecular structure of (**125**)

Figure 4.32   The chemical diagram and molecular structure of (**126**)

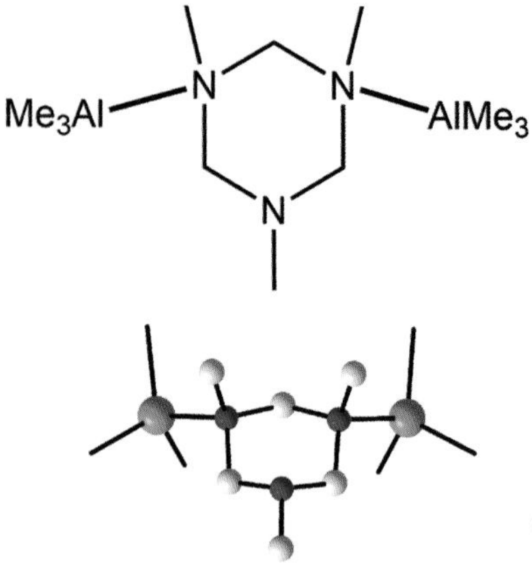

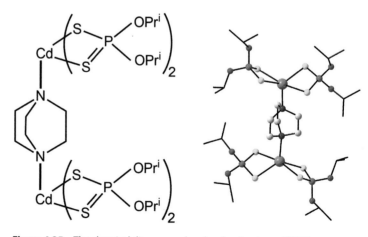

Figure 4.33 The chemical diagram and molecular structure of (**127**)

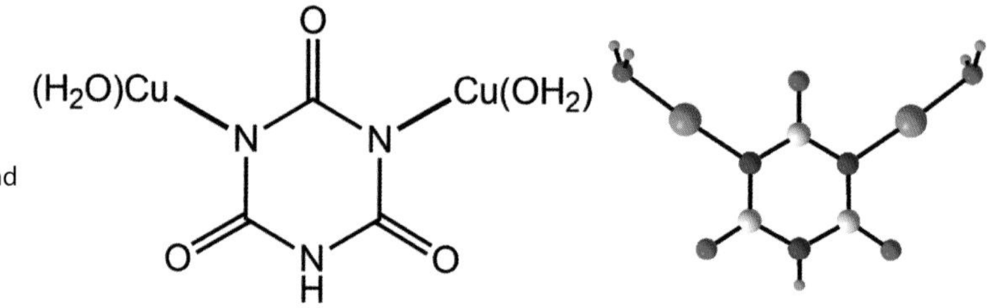

A binuclear compound is constructed in (μ_2-trimethyl-1,3,5-triazinane)[AlMe$_3$]$_2$ (**127**; Figure 4.33) [178], whereby only two of the three potential donors of trimethyl-1,3,5-triazinane are connected to aluminium centres. Just as for compound (**126**), the ring in (**127**) has the form of a chair (Figure 4.33). Disubstitution is also found in the isocyanurato dicopper(I) complex (μ_2-isocyanurato)[Cu(OH$_2$)]$_2$ (**128**; Figure 4.34) [179].

Figure 4.34 The chemical diagram and molecular structure of (**128**)

4.3.2 Bicyclic species

The 1,4-diazabicyclo(2.2.2)octane (DABCO) molecule is a commonly encountered ligand for metals and serves as an efficient inverse coordination centre, as exemplified in the binuclear cadmium dithiophosphate compound (μ_2-DABCO) {Cd[S$_2$P(OPr^i)$_2$]$_2$}$_2$ (**129**; Figure 4.35) [180].

Figure 4.35 The chemical diagram and molecular structure of (**129**)

4.3.3 Tricyclic species

This class of inverse coordination centres is best represented by the urotropine (hexamethylenetetramine or tetraazaadamantane) species, which forms a variety of inverse coordination complexes with varying numbers of nitrogen atoms functioning as donors [181]. All four nitrogen atoms connect atoms in the tetranuclear (μ_4-urotropine)[GaMe$_3$]$_4$ (**130**; Figure 4.36) [182]. The structure of trinuclear (μ_3-urotropine)[AlMe$_3$]$_3$ (**131**; Figure 4.37) [182] and binuclear (μ_2-urotropine){Cd[S$_2$P(OBu^s)$_2$]$_2$}$_2$ (**132**; Figure 4.38) [183] serve as examples of three and two heavy atoms being connected to urotropine, respectively.

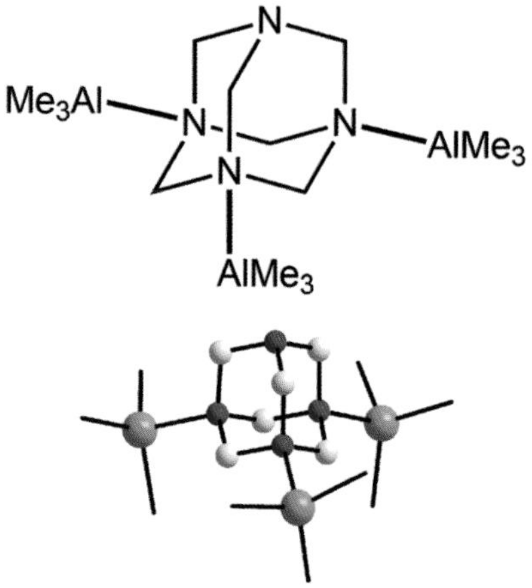

Figure 4.37 The chemical diagram and molecular structure of (**131**)

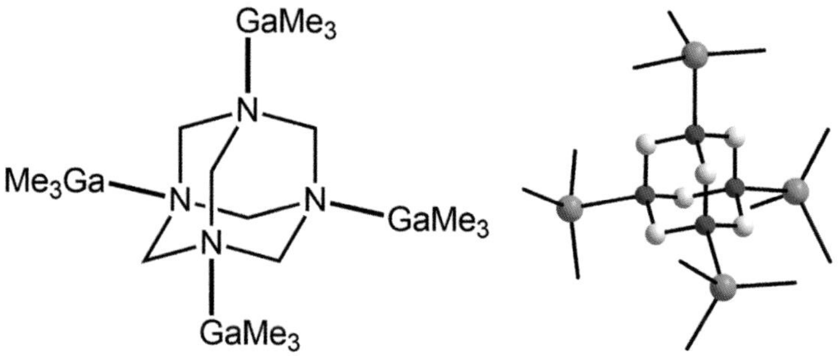

Figure 4.36   The chemical diagram and molecular structure of (**130**)

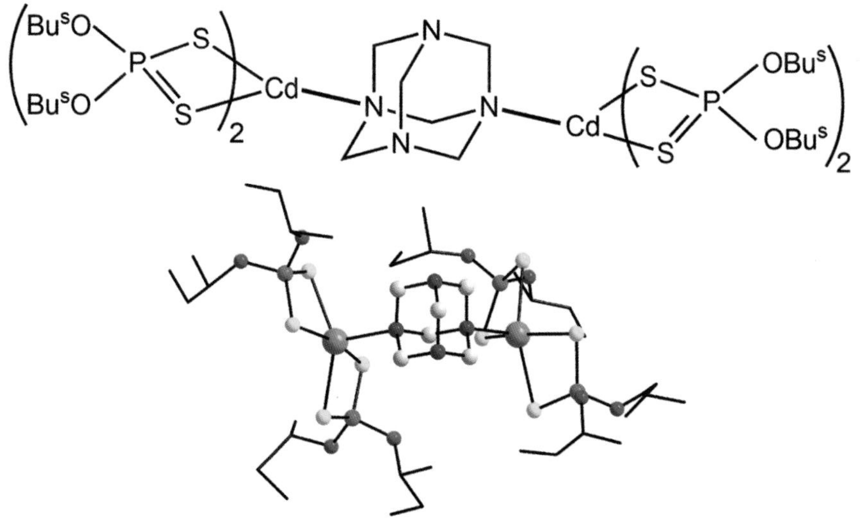

Figure 4.38   The chemical diagram and molecular structure of (**132**)

4.4 Five-membered rings as inverse coordination centres

There are a multitude of five-membered nitrogen-containing heterocycles. These display a great variety of coordination modes, but are found mostly in two- and three-dimensional coordination polymers. Here, the focus turns to zero-dimensional aggregates in which the five-membered ring serves as an inversion coordination centre.

4.4.1 Pyrazole

Pyrazole alone seldom forms inverse coordination complexes as the coordination centre. However, examples do exist, such as in the binuclear platinum cation {(μ_2-pyrazolato)[Pt(NH$_3$)$_2$Cl]$_2$}$^+$ (133; Figure 4.39) [184] and organogold(I) cation, {(pyrazolato)[Au(C$_6$F$_5$)]$_2$}$^+$ (134; Figure 4.40) [185].

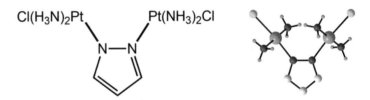

Figure 4.39   The chemical diagram and molecular structure of (**133**)

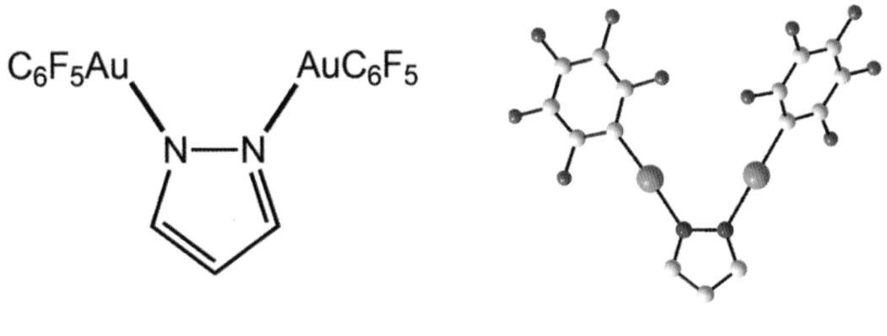

Figure 4.40   The chemical diagram and molecular structure of (**134**)

4.4.2 Imidazole

Imidazole, an isomer of pyrazole, also forms inverse coordination complexes, as exemplified by the structure of the trication {(μ_2-imidazolato)[Cu(2,2′-bipyridyl)$_2$]$_2$}$^{3+}$ (135; Figure 4.41) [186].

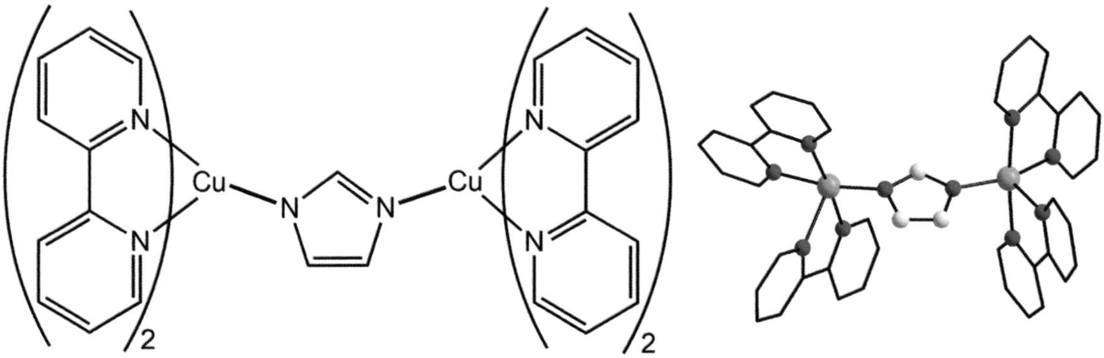

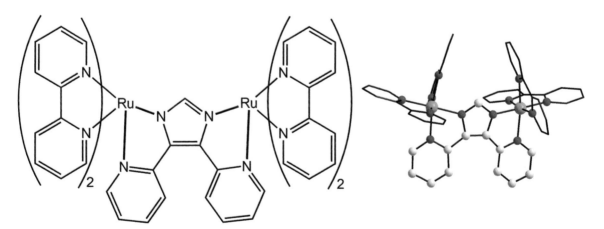

Figure 4.41 The chemical diagram and molecular structure of (**135**)

As aforementioned, the functionalisation of the basic inversion centre, in this case imidazole, enhances the coordination ability and diversifies the metal centres that may be linked. For imidazole, the inclusion of 2-pyridyl groups adjacent to the imidazole-nitrogen donors can lead to the coordination of octahedral ruthenium(II) in tricationic $\{\mu_2$-(2,3-di-2-pyridyl-imidazolato)[Ru(2,2′-bipyridyl)$_2$]$_2\}^{3+}$ (**136**; Figure 4.42) [187].

Figure 4.42 The chemical diagram and molecular structure of (**136**)

In an allied manner but by introducing charged substituents onto the imidazole ring (that is, carboxylato moieties), the binuclear silver(I) species, {(3,5-dicarboxylato-pyrazolato)[Ag(NH₃)]₂} (137; Figure 4.43) [188], can be formed. Indeed, this is a rare example of a mixed substitution pattern in an inverse coordination centre complex as the structure contains carboxylic acid/carboxylato residues in the overall neutral molecule. The carboxylic acid/carboxylato residues are self-associated via an intramolecular carboxylic acid–O–H⋯O(carboxylato) hydrogen bond.

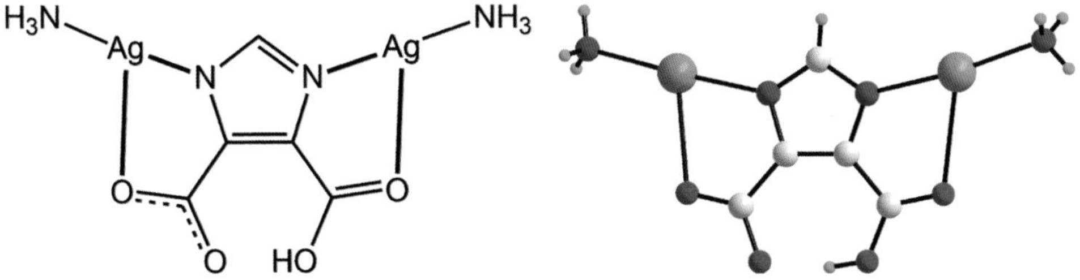

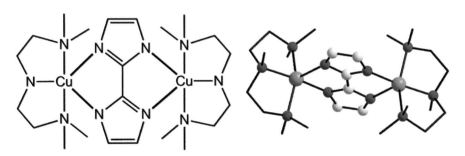

Figure 4.43   The chemical diagram and molecular structure of (**137**)

4.4.3 Biimidazole

As will be demonstrated, biimidazole is also a versatile inverse coordination centre. Two of its complexes with copper(II), in the dication, {(μ₂-bi-imidazolato)[Cu(pentamethyldiethylenetriamine)]₂}²⁺ (**138**; Figure 4.44) [189], and in the heterometallic dianion, {(μ₂-bi-imidazolato)[Sn(η¹-Cr(CO)₅)₂]₂}²⁻ (**139**; Figure 4.45) [190], are illustrated here.

Figure 4.44   The chemical diagram and molecular structure of (**138**)

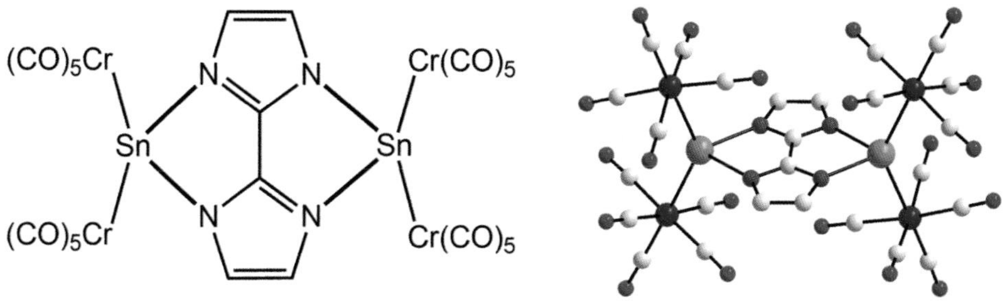

Figure 4.45 The chemical diagram and molecular structure of (**139**)

As seen earlier, appending aryl rings to the basic framework of the biimidazole inverse coordination centre may lead to similar architectures, as in (μ_2-2,2'-bibenzimidazolato)[Re(CO)$_3$(pyridine)]$_2$ (**140**; Figure 4.46) [191], where each rhenium(I) centre is chelated, and examples with fewer metal centres connected, as in (μ_2-2,2'-bibenzimidazolato)[Au(PPh$_3$)]$_2$ (**141**; Figure 4.47) [192], with linearly coordinated gold(I) centres. In centrosymmetric compound (**141**), it is noted that the gold(I) atom is coplanar with the imidazole ring and is in close proximity to the non-coordinating nitrogen atom. However, the Au···N separation of 3.00 Å is not indicative of a significant bonding interaction.

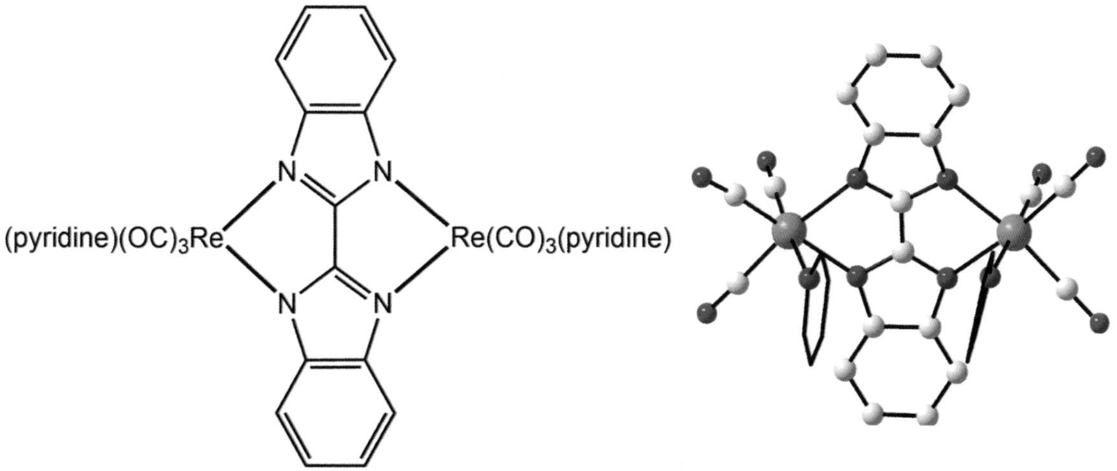

Figure 4.46 The chemical diagram and molecular structure of (**140**)

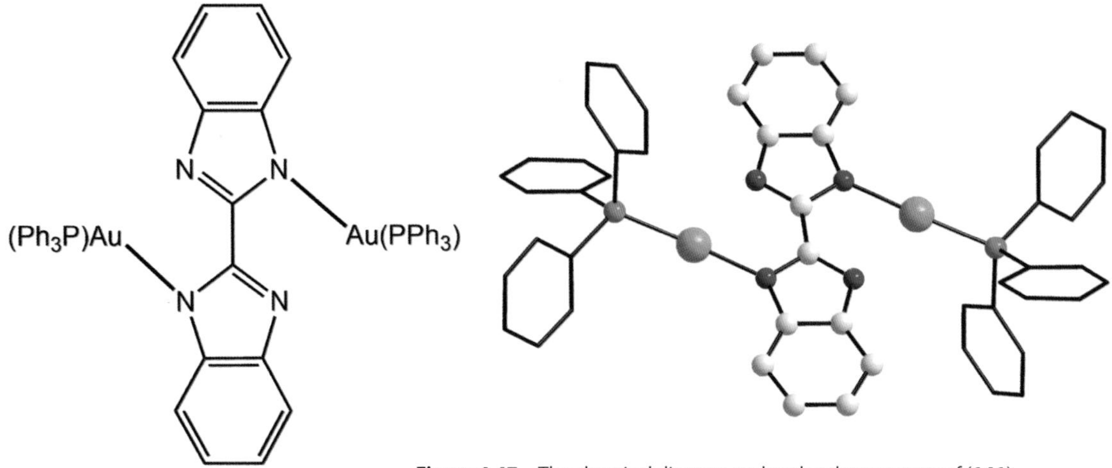

Figure 4.47 The chemical diagram and molecular structure of (**141**)

Indeed, it is possible to attach two additional Au(PPh$_3$) residues to give the {(μ_4-2,2′-bi-benzimidazolato)[Au(PPh$_3$)]$_4$}$^{2+}$ dication (**142**; Figure 4.48) [192]. Here, the close proximity of the gold(I) centres enables the formation of weak aurophilic (Au···Au) interactions of 3.16 and 3.22 Å (not shown in the figure). The inversion coordination centre in the non-symmetric compound (**142**) is significantly twisted compared with those observed in the earlier examples, in this case, presumably to minimise steric repulsion between the Au(PPh$_3$) residues.

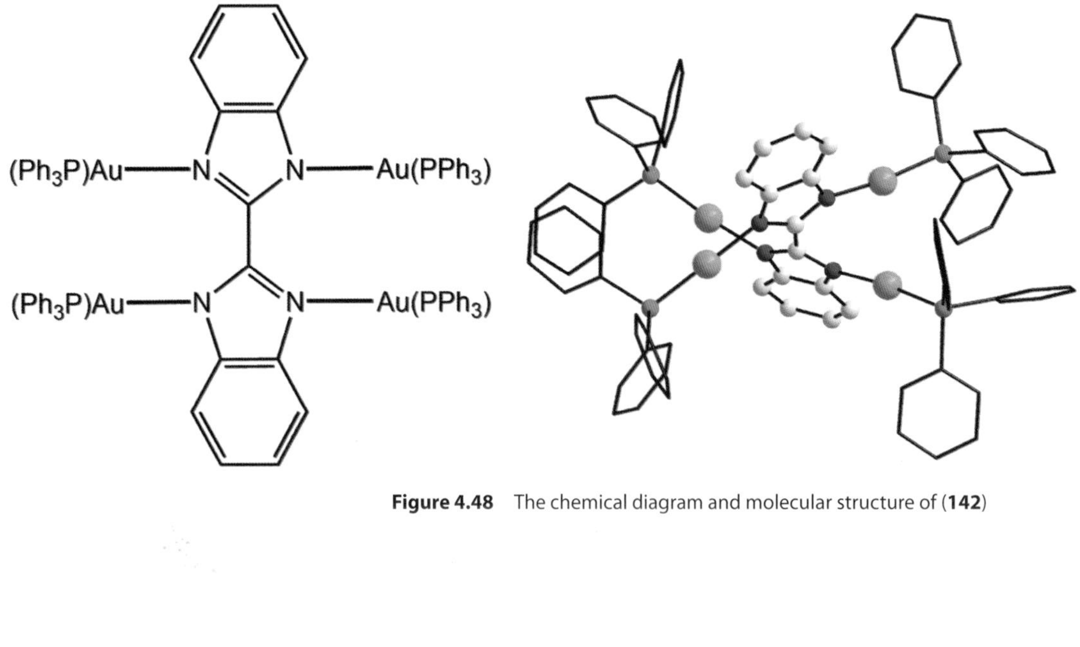

Figure 4.48 The chemical diagram and molecular structure of (**142**)

In a variation of a theme, one of the imidazole rings of the biimidazole inverse coordination centre can be replaced by a 2-pyridyl ring to yield the {(μ$_2$-2-(pyridin-2-yl)-benzimidazolato)[Au(PPh$_3$)]$_2$}$^+$ monocation (**143**; Figure 4.49) [193], in which case the two Au(PPh$_3$) residues are connected to the five-membered ring exclusively. The close approach of the pyridyl-nitrogen atom (2.69 Å) is noted but, as both P–Au–N angles are 174°, this interaction is not significant.

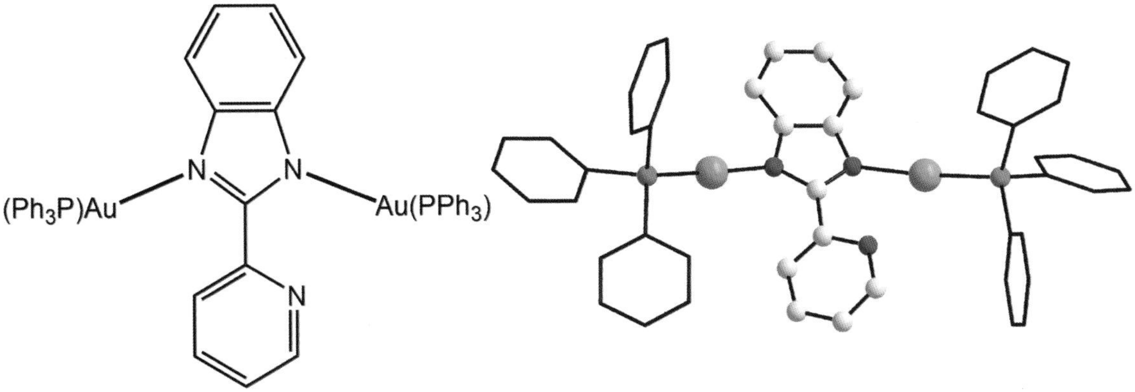

Figure 4.49   The chemical diagram and molecular structure of (**143**)

4.4.4 Triazole

In addition to numerous coordination polymers and metal-organic frameworks constructed about triazole and derivatives [194], a few discrete molecular inverse coordination complexes have been described. These include the monoanionic dirhodium(I) complex of 1,2,3-triazole-4,5-dicarboxylate, {(μ$_2$-4,5-dicarboxytriazolato)[Rh(CO)$_2$]$_2$}$^-$ (**144**; Figure 4.50) [195], whereby the carboxylato-oxygen atoms are coordinating.

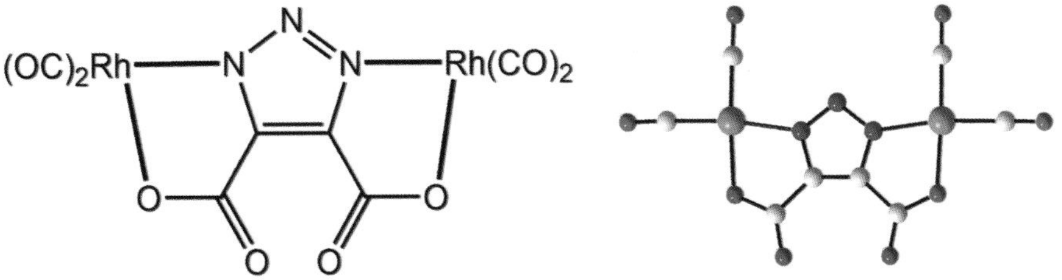

Figure 4.50   The chemical diagram and molecular structure of (**144**)

Another interesting example is found in the substituted benztriazole compound, {(μ₂-2H-benzotriazo-2-lato)-6-((diethylamino)methyl)-4-methylphenolato-dimethylaluminium}-[AlMe₃]₂ (145; Figure 4.51) [196]. Here, the triazole functions as an inverse coordination centre for two AlMe₃ residues while the pendant phenolate and diethylamine groups coordinate a dimethylaluminium centre.

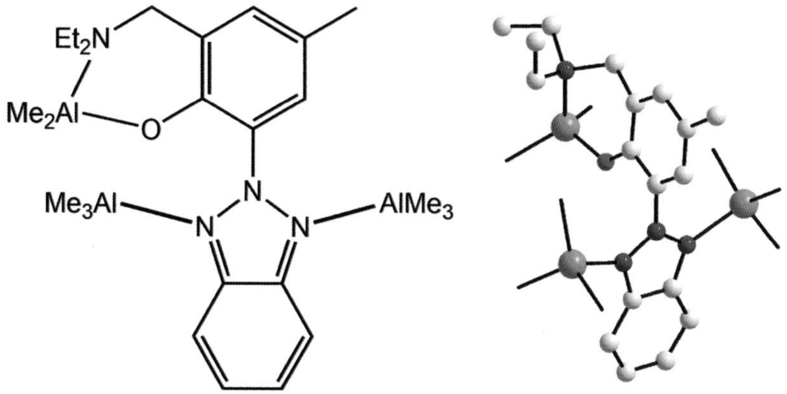

Figure 4.51   The chemical diagram and molecular structure of (**145**)

Inverse coordination complexes with 1,2,4-triazole are rare. An example is illustrated by the copper(I) dication {(5-t-butyl-1,2,4-triazo-3-lato)pyrimidine)[Cu(PPh₃)₂]₂}²⁺ (146; Figure 4.52) [197], in which an appended pyrimidine ring cooperates with the triazole ring nitrogen atoms to chelate the copper(I) centres.

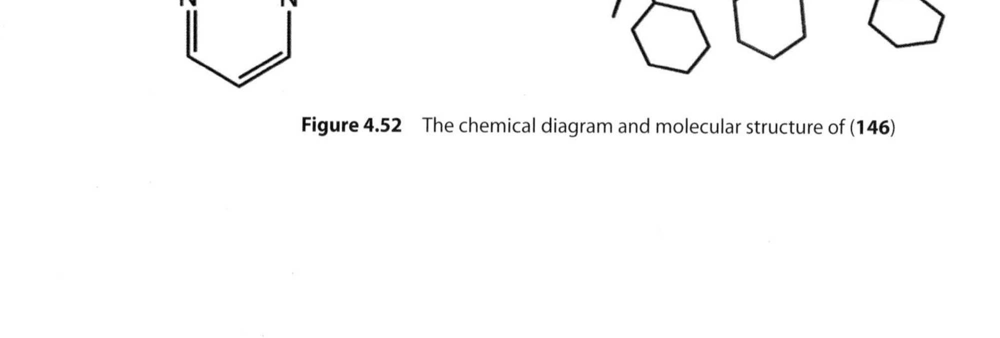

Figure 4.52   The chemical diagram and molecular structure of (**146**)

4.4.5 Tetrazole

Tetrazole is a versatile polytopic ligand mostly involved in the formation of coordination polymers and metal-organic frameworks [198]. Thus, only a few discrete molecular inverse coordination complexes have been described. These include the binuclear rhodium(III) tricationic species, {(μ_2-5-methyltetrazolato) [Rh(Cp*)(2,2'-bipyridyl)]$_2$}$^{3+}$ (**147**; Figure 4.53) [199], and, resembling compound (**146**), the binuclear nickel(II) trication, (μ_2-5-(2-pyrimidyl)tetrazolato)[Ni(NH$_2$CH$_2$CH$_2$)$_3$N]$_2$}$^{3+}$ (**148**; Figure 4.54) [200].

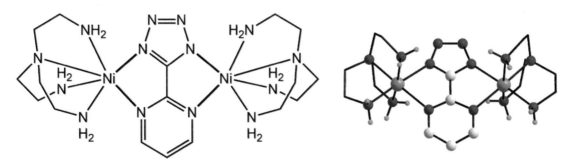

Figure 4.53   The chemical diagram and molecular structure of (**147**)

Figure 4.54   The chemical diagram and molecular structure of (**148**)

4.4.6 Fused five-membered rings

Two inverse coordination gold(I) phosphane complexes built around fused five-membered rings serve as exemplars for this class of complexes. These are (μ_2-3,6-bis(4-chlorophenyl)-2,5-dihydropyrrolo(3,4-c)pyrrole-1,4-dione-2,5-diyl)[Au(PPh$_3$)]$_2$

(**149**; Figure 4.55) [201] and (μ_2-3,6-bis(4-pyridyl)-2,5-dihydropyrrolo(3,4-c)pyrrole-1,4-dione)[Au(PPh$_3$)]$_2$ (**150**; Figure 4.56) [202]. With pendent pyridyl groups, the structure of compound (**150**) offers opportunities for further complexation and the generation of heterometallic species.

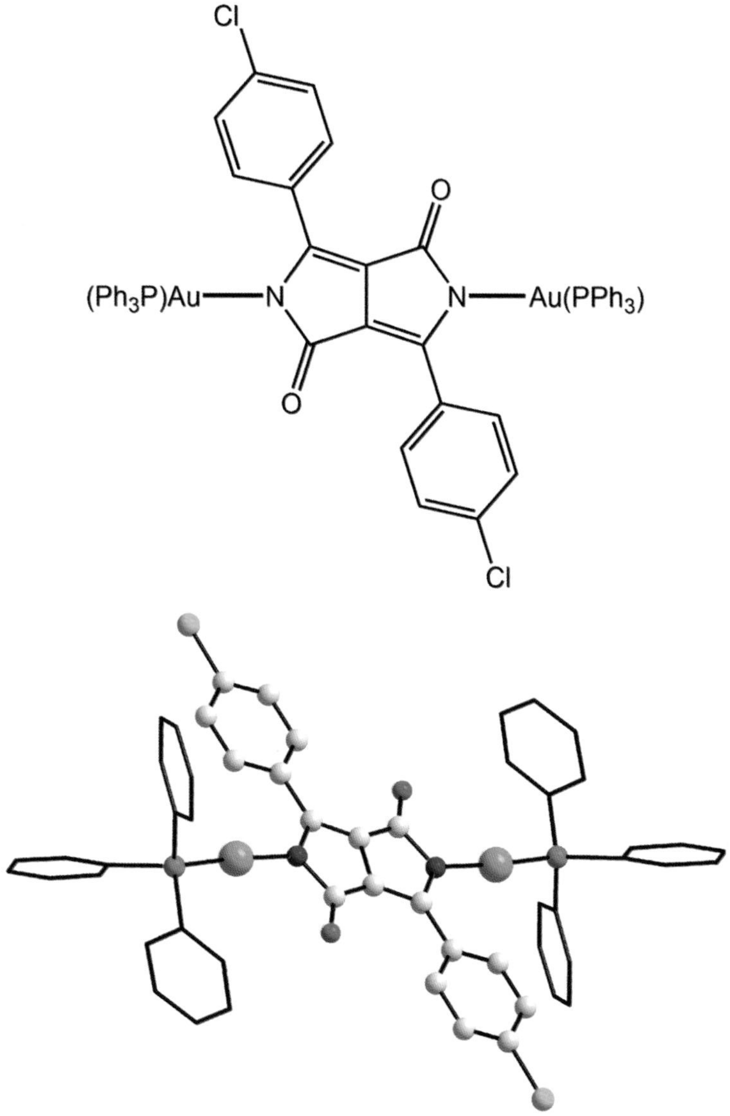

Figure 4.55 The chemical diagram and molecular structure of (**149**)

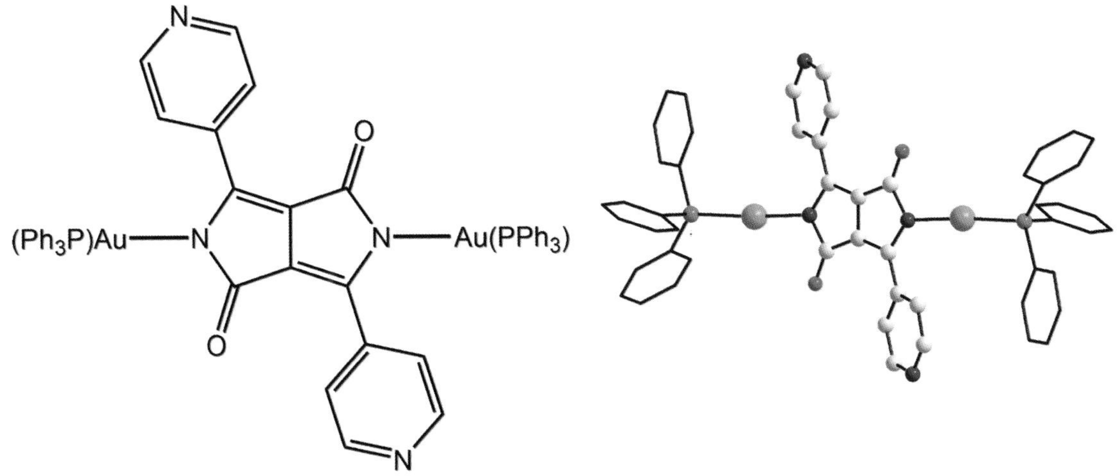

Figure 4.56 The chemical diagram and molecular structure of (**150**)

4.5 Carboxylates as inverse coordination centres

Carboxylato anions are excellent coordination ligands and powerful instruments for robust connection to metal centres. A wide variety of coordination modes of carboxylates are known, and these often lead to the stabilisation of two- and three-dimensional coordination polymers [203]. When two or more carboxylato functions are present in a molecule, the resultant polytopic ligand may become the centre of an inverse coordination complex. The best centres are the aromatic carboxylato anions, but aliphatic carboxylic acids can also be useful in this context.

Aromatic dicarboxylates, with the COO⁻ function in mutual *ortho*-positions, are not very frequently encountered as inverse coordination centres, an observation probably related to steric reasons. However, examples are known, for instance, in the binuclear silver complex [(μ$_2$-phthalato)(AgNH$_3$)$_2$] (**151**; Figure 4.57) [204].

Para-dicarboxylic ligands with bulky terminal groups that prevent polymer formation are the most common form of inversion coordination complex of this class of compounds. Some examples include the dicationic, octahedral cobalt(II) complex {(μ$_2$-terephthalato)[Co(2,2′-bipyridine)$_2$]$_2$}$^{2+}$ (**152**; Figure 4.58) [205], in which the carboxylato groups adopt a chelating mode.

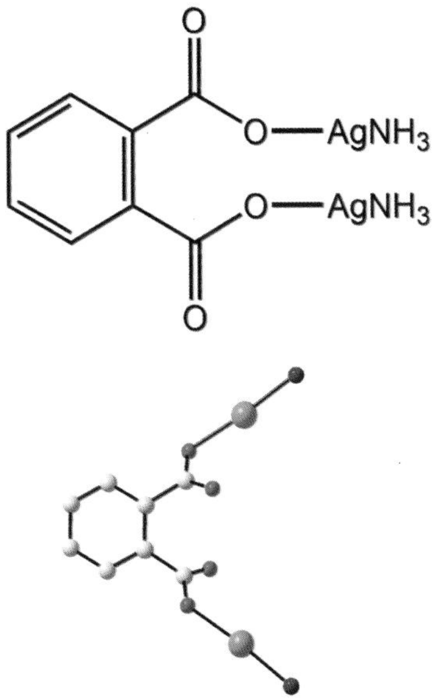

Figure 4.57 The chemical diagram and molecular structure of (**151**) (note that ammonia–hydrogen atoms are not shown)

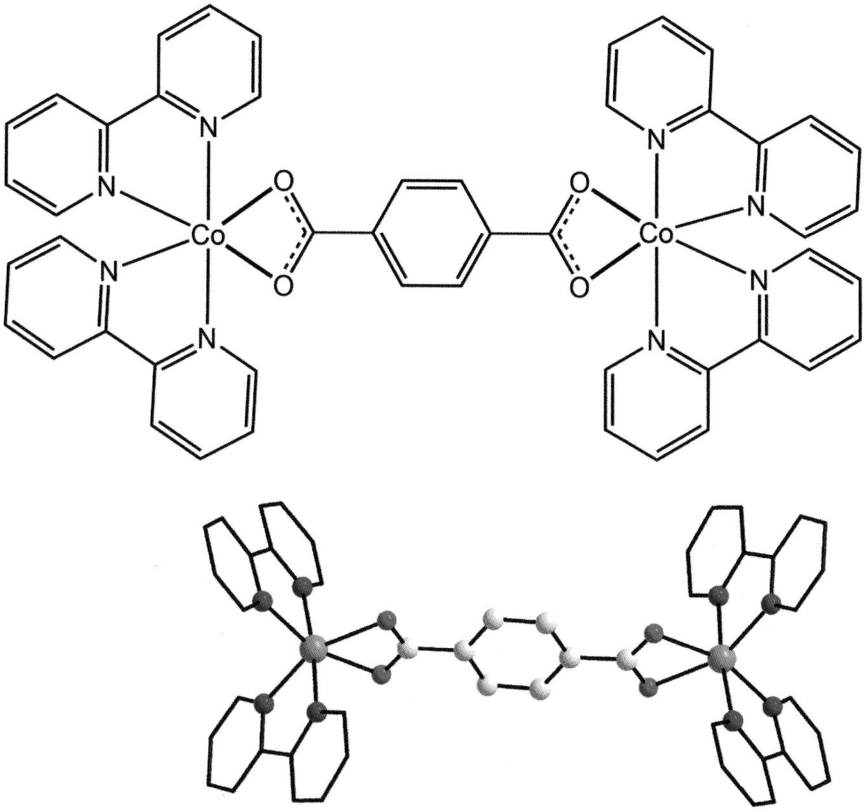

Figure 4.58   The chemical diagram and molecular structure of (**152**)

Another example is found in the binuclear octahedral nickel(II) inverse coordination complex, (μ_2-benzene-1,2,4,5-tetracarboxylato)[Ni(2,2′-bipyridine)(OH$_2$)$_3$]$_2$ (**153**; Figure 4.59) [206], with single points of contact between the inverse coordination centre and each metal centre.

Inverse coordination complexes with the two active carboxylato functions in *meta*-positions are also known, such as in the organotin compound (μ_2-4,6-bis(4-methylbenzoyl)isophthalato)[SnPh$_3$(OH)]$_2$ (**154**; Figure 4.60) [207]. The carboxylato ligands coordinate the tin atoms in a monodentate mode leading to *trans*-C$_3$O$_2$ trigonal-bipyramidal coordination geometries.

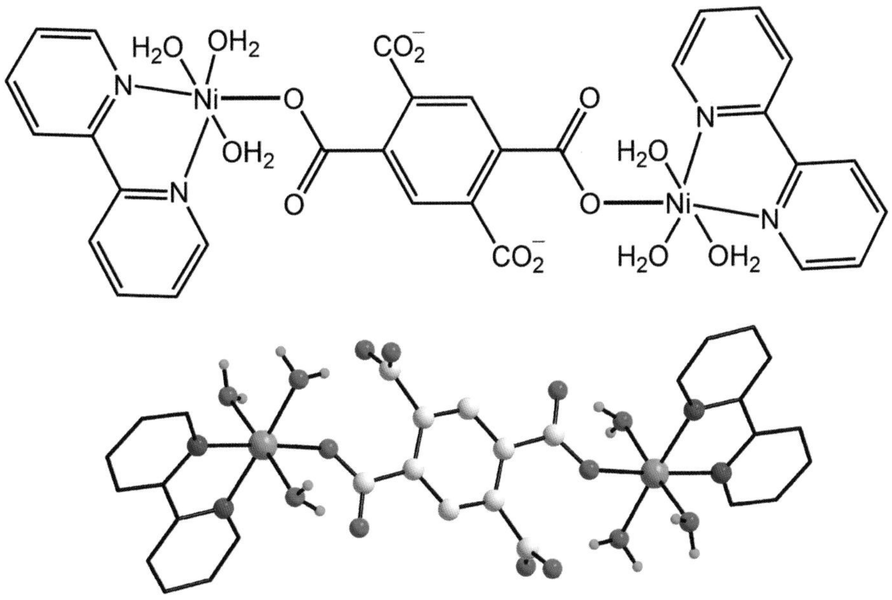

Figure 4.59 The chemical diagram and molecular structure of (**153**)

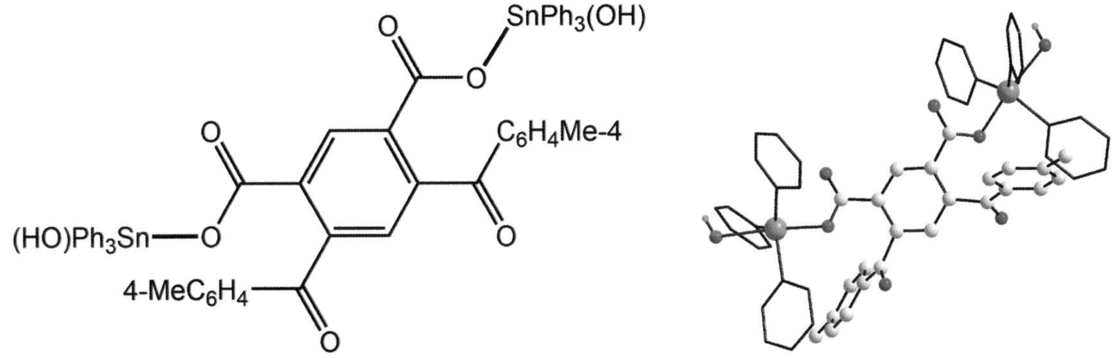

Figure 4.60 The chemical diagram and molecular structure of (**154**)

It is also possible to increase the number of carboxylato substituents to generate higher nuclearity organotin inverse coordination complexes. Two illustrated examples are formed with benzene 1,3,5-tricarboxylato and 1,2,4,6-tetracarboxylato leading, respectively, to trinuclear (μ_3-benzene-1,3,5-tricarboxylato) [SnPh$_3$(OH)]$_3$ (**155**; Figure 4.61) [208] and tetranuclear (μ_4-benzene-1,2,4,5-tetracarboxylato)[SnPh$_3$(DMF)]$_3$ (**156**; Figure 4.62) [209]. The tin atom in compound (**155**) exists within a *cis*-C$_3$O$_2$

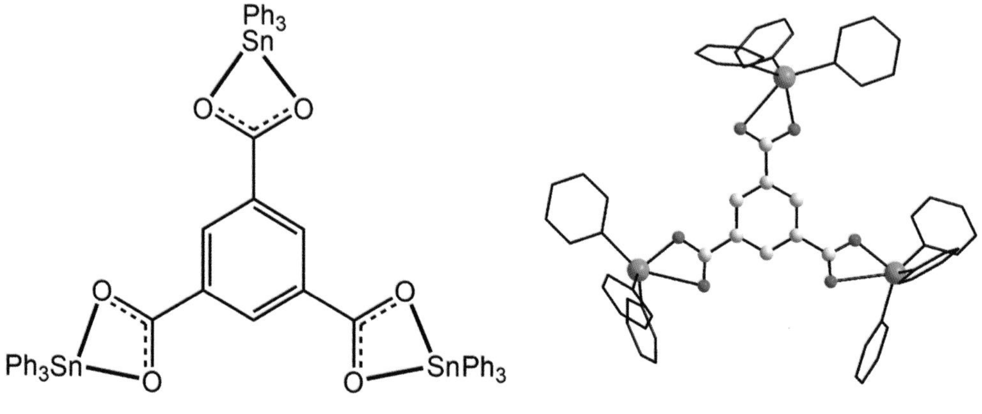

Figure 4.61   The chemical diagram and molecular structure of (**155**)

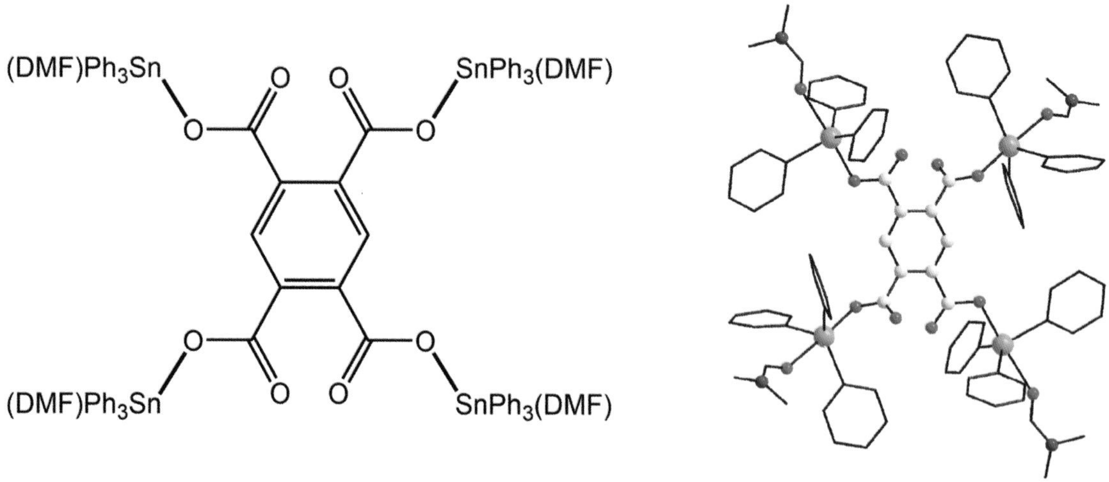

Figure 4.62   The chemical diagram and molecular structure of (**156**)

trigonal-bipyramidal environment as the carboxylato ligand is chelating. By contrast, in compound (156), *trans*-C_3O_2 trigonal-bipyramidal coordination geometries are found, as in compound (154), a result likely to be correlated with steric congestion in these molecules.

As a final example in this category, a rare mode of inverse coordination is orchestrated by the benzene-1,2,4,5-tetracarboxylato inverse coordination centre, also seen in compound (156). Here, adjacent carboxylato ligands employ one oxygen atom each to coordinate octahedral nickel(II) centres, leading to seven-membered chelate rings, in the structure of (μ_2-benzene-1,2,4,5-tetracarboxylato)[Ni(1,10-phenanthroline) (OH$_2$)]$_2$ (157; Figure 4.63) [210].

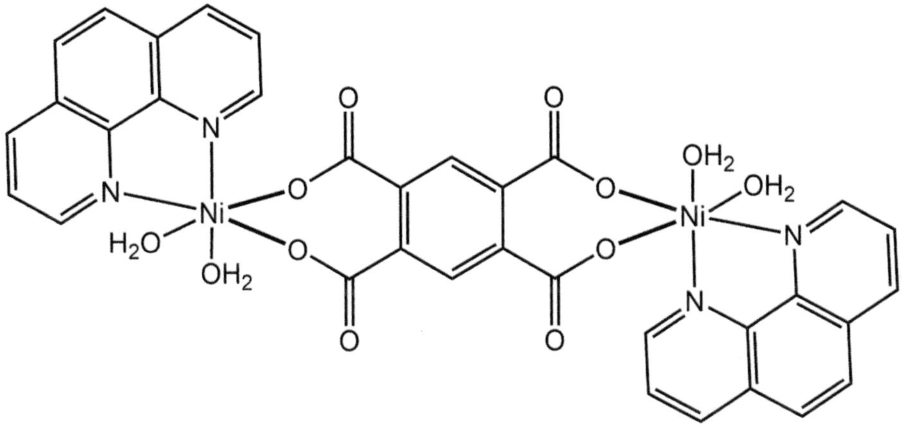

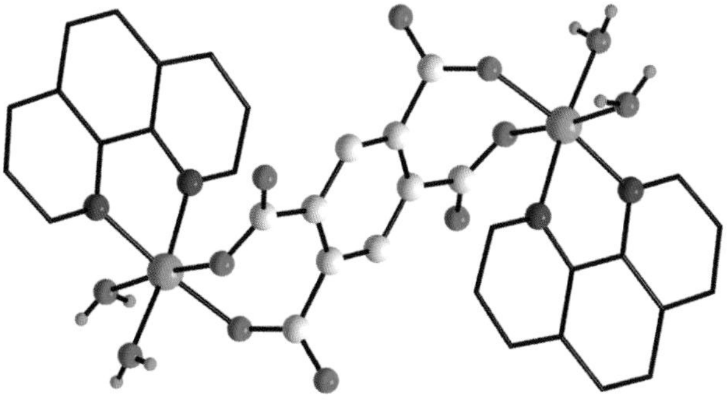

Figure 4.63   The chemical diagram and molecular structure of (157)

4.6 Oxocarbons as inverse coordination centres

Oxocarbon anions comprise an interesting class of ligands as, owing to their inherently polytopic character, they are well suited for the formation of inverse coordination complexes [211]. Oxocarbon anion-based inverse coordination complexes with a three-membered ring are rare, perhaps owing to putative steric congestion. An example with considerable aromatic character in the inversion centre is found in the binuclear uranium(IV) complex, (μ_2-deltato)[U(η^8-1,4-bis(tri-isopropylsilyl)cyclo-octatetraene)(Cp*)]$_2$ (**158**; Figure 4.64) [212]. The illustrated images in Figure 4.64 are a simplification as there are also U$\cdots$C(deltato) interactions contributing to the stability of the complex [212].

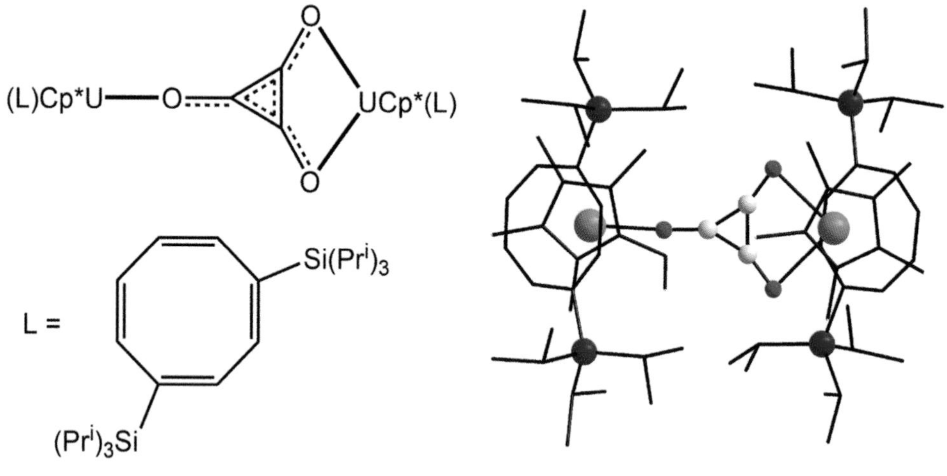

Figure 4.64   The chemical diagram and molecular structure of (**158**)

In another example, (μ_3-cyclopropane-1,2,3-triolato){Mg[bis(2,6-diethylphenyl)pentane-2,4-di-iminato)]}$_3$ (**159**; Figure 4.65) [213], the carbon atoms comprising the inverse coordination centre are sp^3-hybridised and the three magnesium(II) substituents lie to one side of the ring, and each can be considered chelated by adjacent oxygen atoms leading to a Mg_3O_3 ring.

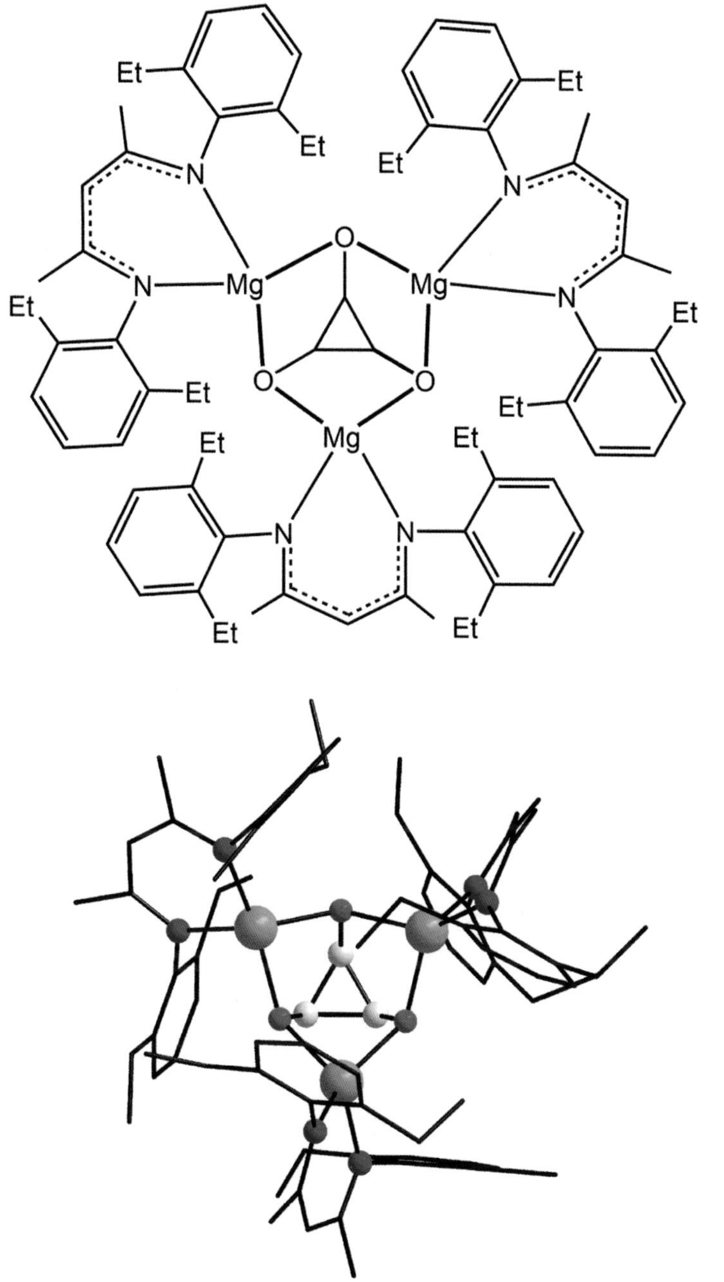

Figure 4.65 The chemical diagram and molecular structure of (**159**)

While the bonding remains essentially the same in the following examples where the inverse coordination centre is based on the squarato oxocarbon ring, different arrangements arise depending on the positions of attachment. In the copper(II) dication $\{(\mu_2\text{-squarato})[Cu(1,10\text{-phenanthroline})_2]\}^{2+}$ (**160**; Figure 4.66) [214], the inverse coordination centre functions as a dianion and coordination occurs in 1,2-positions about the ring.

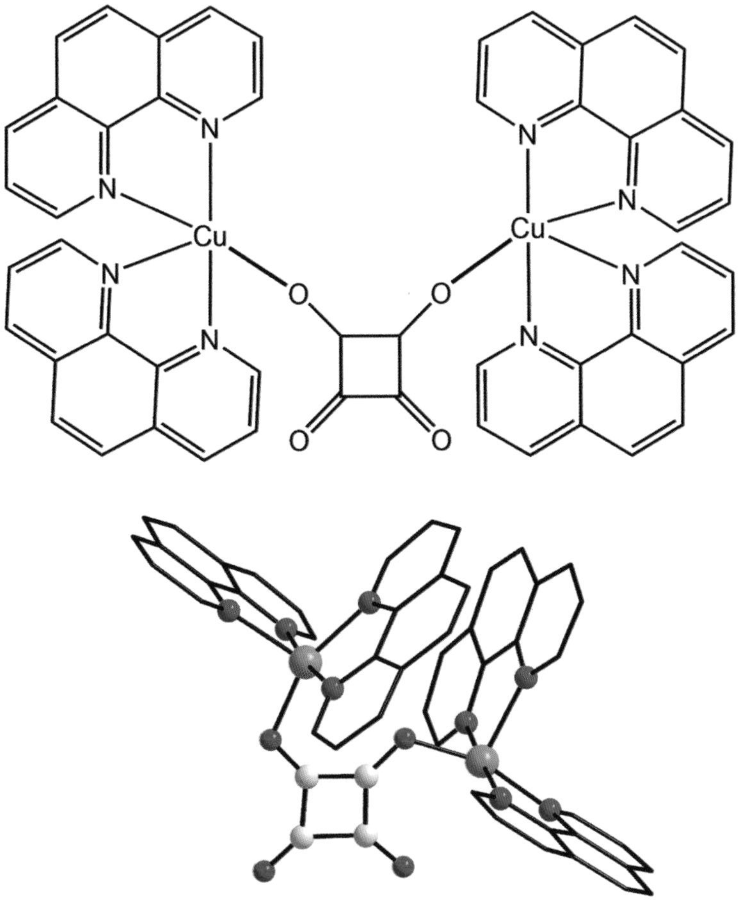

Figure 4.66   The chemical diagram and molecular structure of (**160**)

In the second squarato example, namely (μ_2-squarato)[Cu(2,2'-bipyridyl)Br(OH$_2$)]$_2$ (**161**; Figure 4.67) [215], the copper(II) centres—within approximate square-pyramidal geometries—are diagonally opposite.

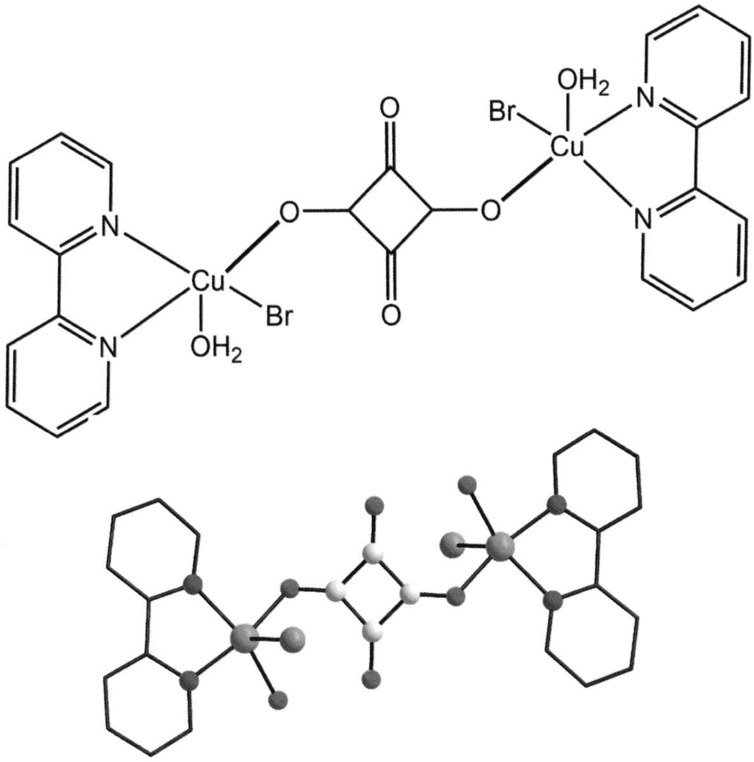

Figure 4.67 The chemical diagram and molecular structure of (**161**) (note that water-hydrogen atoms are not shown)

A very similar function for the planar croconato dianion, as the inverse coordination centre, is evident in the molecular structure of (μ_2-croconato){Cu[(bis(2-pyridylcarbonyl)amido]$_2$}$_2$ (**162**; Figure 4.68) [216], in which the copper(II) species exists within a 4+1 coordination geometry. The weaker interactions with croconato-carbonyl oxygen atoms, at 2.48 and 2.58 Å (the molecule lacks symmetry), compared to the formal Cu–O bond lengths of 1.96 and 1.97 Å are not shown in Figure 4.68.

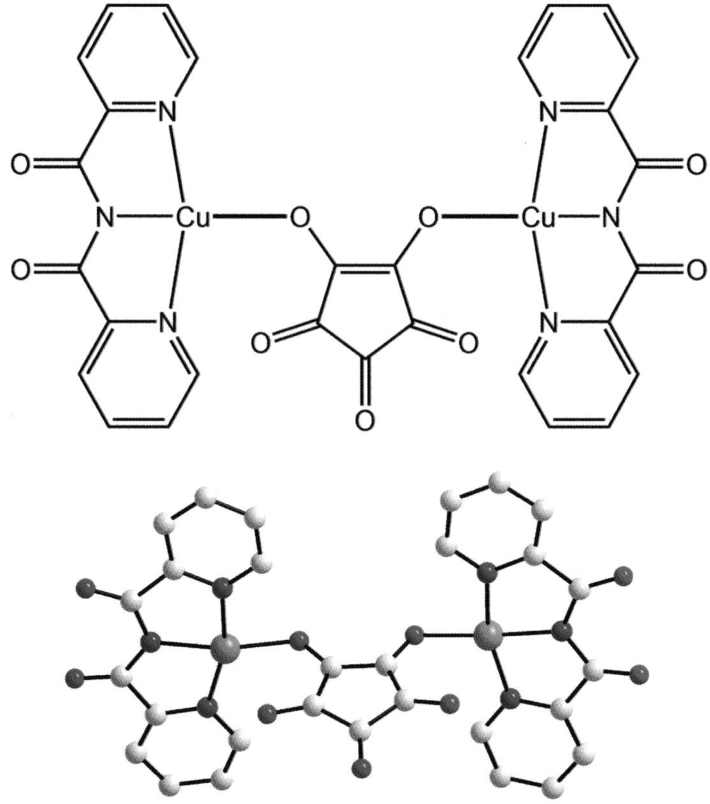

Figure 4.68   The chemical diagram and molecular structure of (**162**)

An example of a hexolato inverse coordination centre is found in (μ_6-hexolato)[Re(CO)$_3$(pyridine)]$_6$ (**163**; Figure 4.69) [217]. Here, six rhenium(I) centres decorate the central ring, each being chelated by adjacent oxygen atoms leading to octahedral coordination and an attractive peripheral Re$_6$O$_3$ star arrangement, which is planar as the ring is generated by $\bar{3}$ crystallographic symmetry.

Related inverse coordination complexes can be generated by judicious substitution, such as that derived from chloranilic acid, which combines halogeno and oxo functionalities. An example is found in the copper(II) dication {(μ_2-chloranilic acid) [Cu(2,2′:6′,2″-terpyridine)]$_2$}$^{2+}$ (**164**; Figure 4.70) [218]. Here, the differential between the Cu–O bonds is not as great as in, for example structure (**162**), that is 1.97 and 2.27 Å (centrosymmetric molecule) so the copper(II) can be thought as being chelated by the oxygen atoms leading to a square-pyramidal geometry.

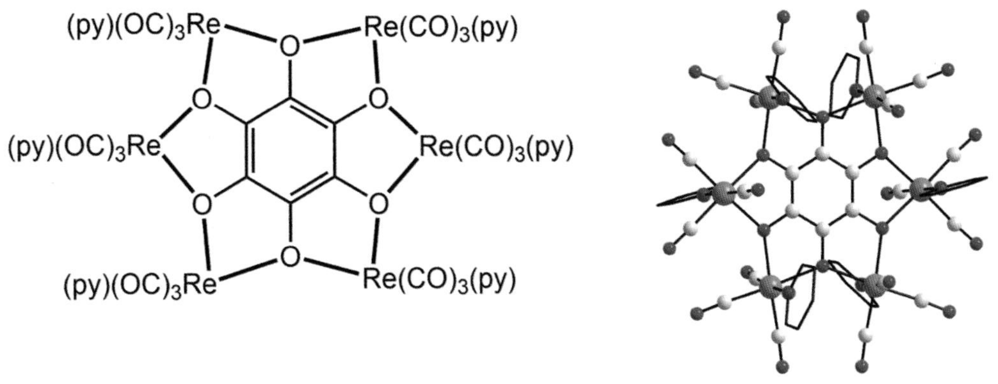

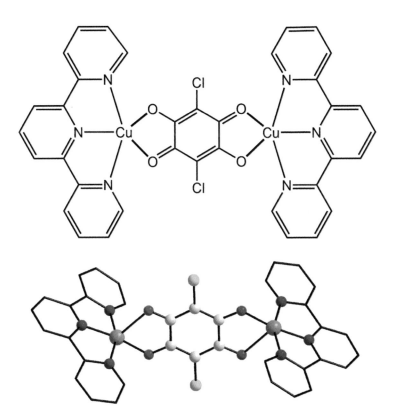

Figure 4.69   The chemical diagram and molecular structure of (**163**)

Figure 4.70   The chemical diagram and molecular structure of (**164**)

4.7 Oxalate as an inverse coordination centre, and nitrogen and sulphur analogues

Oxalate and products of partial or complete replacement of oxygen with nitrogen and sulphur are very versatile ligands for inverse coordination, normally by forming two fused chelate rings. A prototype example with oxalate as the core is seen in binuclear (μ_2-oxalato)[Cu(2,2′-bipyridine)(THF)]$_2$ (**165**; Figure 4.71) [219]. Other symmetric species are obtained when both oxygen atoms are replaced by N(Me), as in the organoaluminium compound, (tetramethyl-oxalamidinato)[AlMe$_2$]$_2$ (**166**; Figure 4.72) [220], or by N(Ph), as in the binuclear nickel(II) complex, (μ_2-bis(diphenyl) oxalamidinato)[Ni(acetylacetonato)]$_2$ (**167**; Figure 4.73) [221].

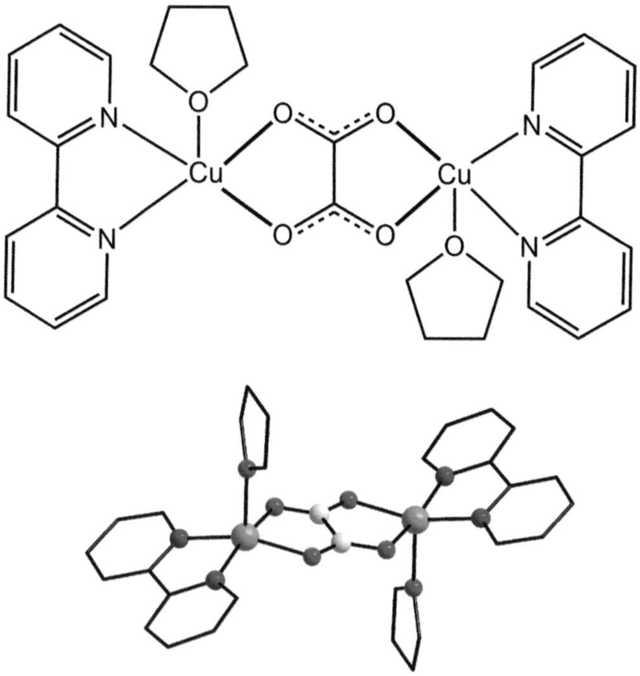

Figure 4.71   The chemical diagram and molecular structure of (**165**)

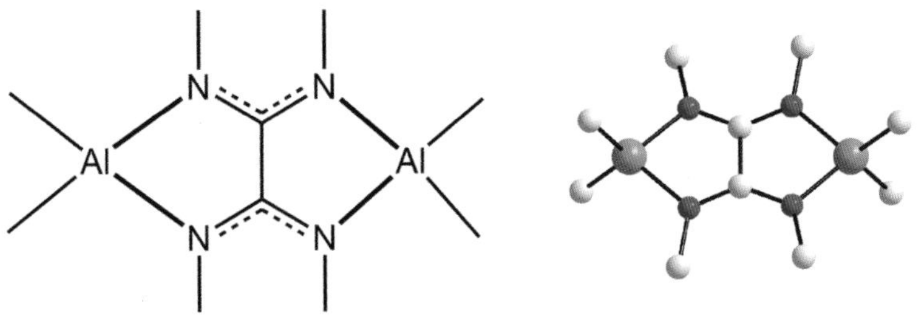

Figure 4.72   The chemical diagram and molecular structure of (**166**)

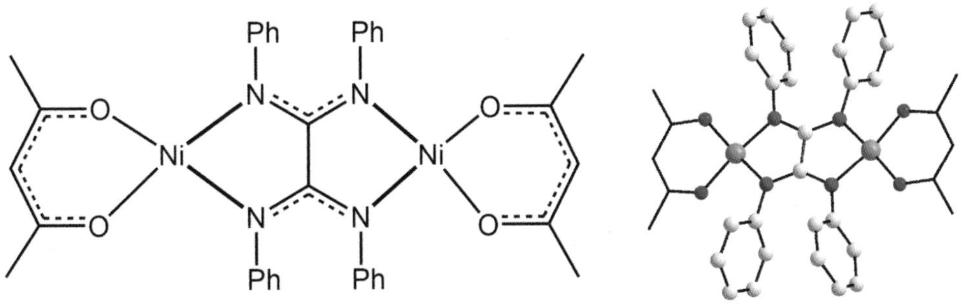

Figure 4.73   The chemical diagram and molecular structure of (**167**)

The substitution of the oxygen atoms by sulphur gives rise to thiooxalato inversion coordination centres, such as in the binuclear rhodium(II) species (μ_2-tetrathio-oxalato)[Rh(Cp*)]$_2$ (**168**; Figure 4.74) [222]. More elaborate examples are also known, for example, in dicationic (μ_2-ethanebis(dithioato))[Ni(1,4-dihydropyrazine-2,3-dithiolato)]$_2$}$^{2+}$ (**169**; Figure 4.75) [223].

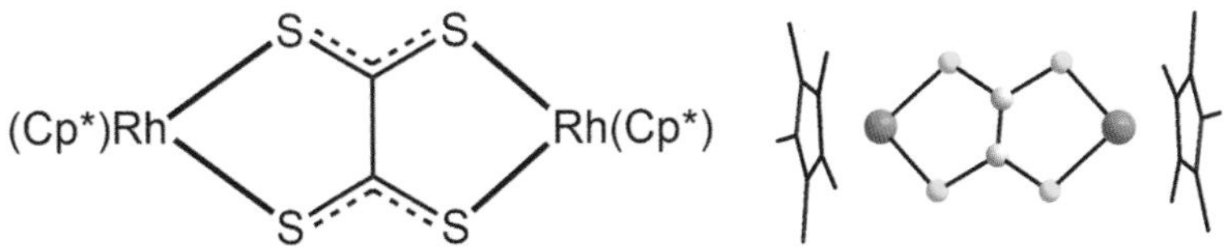

Figure 4.74   The chemical diagram and molecular structure of (**168**)

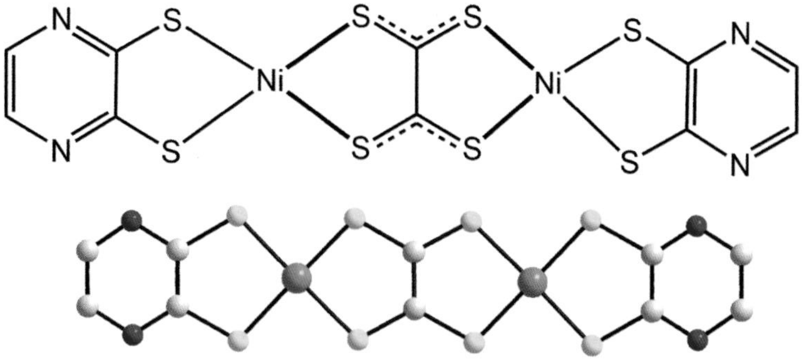

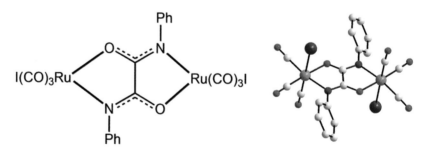

Figure 4.75 The chemical diagram and molecular structure of (**169**)

Partial replacement of the oxygen atoms by other donors are also possible. All subsequent illustrated examples have one hetero group at each carbon atom and always a *trans/anti* relationship across the C–C bond. Substitution of oxygen with N(Ph) occurs in (μ_2-diphenylethanediamidato)[Ru(CO)$_3$I]$_2$ (**170**; Figure 4.76) [224]. Alternatively, the nitrogen can be substituted with additional donor atoms, such as in the case of (μ_2-oxamide-diacetato) [Ni(OH$_2$)$_3$]$_2$ (**171**; Figure 4.77) [225], where a carboxylato group forms an additional chelate for the octahedral nickel(II) atom.

Partial substitution by sulphur is found in tetraanionic (μ_2-1,2-dithio-oxalato)[In(1,2-dithio-oxalato)]$_2$}$^{4-}$ (**172**; Figure 4.78) [226], where the indium(III) centres are within octahedral OS$_5$ geometries.

Figure 4.76 The chemical diagram and molecular structure of (**170**)

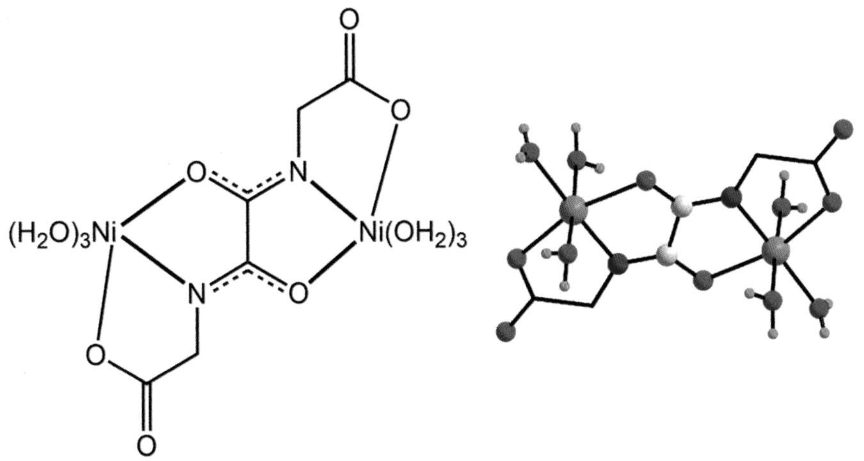

Figure 4.77 The chemical diagram and molecular structure of (**171**)

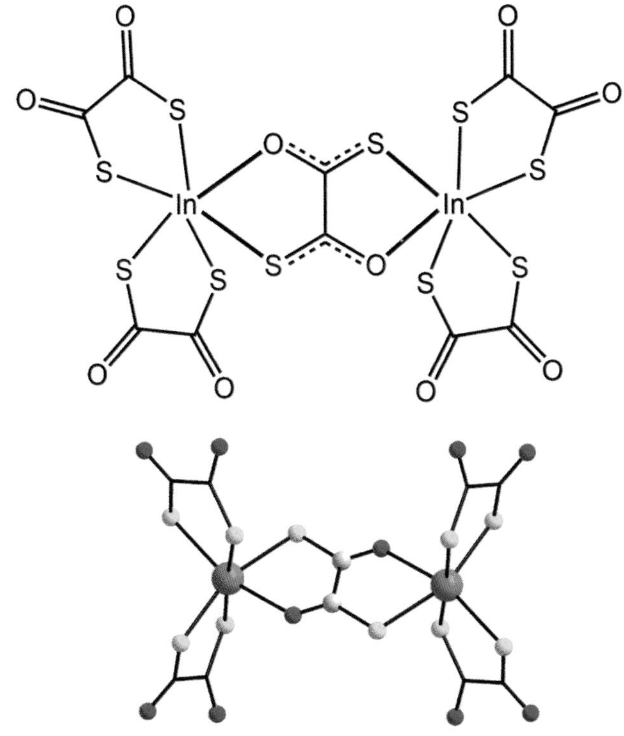

Figure 4.78 The chemical diagram and molecular structure of (**172**)

Mixed combinations of oxalate-type inverse coordination centres, with *cis/syn* sulphur and nitrogen donors have been observed to give rise to non-symmetric species, for example in (μ_2-bis(1-phenylethyl)ethanebis(imidothioato)[(allyl)Pd][Pd(PPr^n)$_3$Cl] (**173**; Figure 4.79) [227].

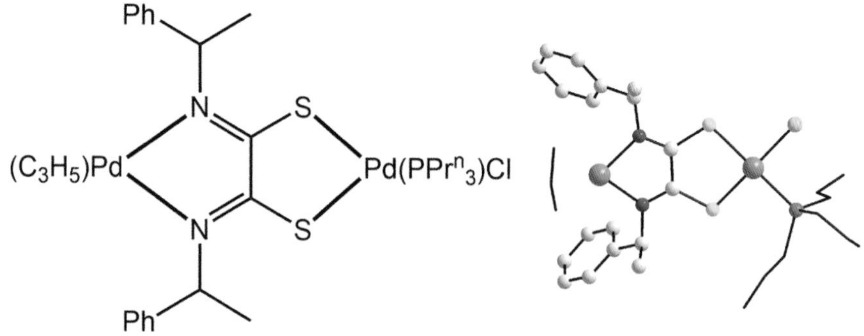

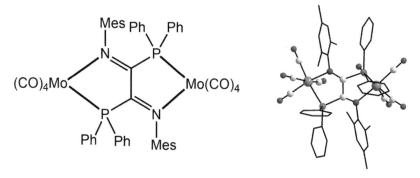

Figure 4.79   The chemical diagram and molecular structure of (**173**)

Finally, for this category of inverse coordination centres, more exotic derivatives are available, including the mixed nitrogen-phosphorus species observed in (μ_2-dimesityl-2,3-bis(diphenylphosphino)-1,4-diazabuta-1,3-diene)[Mo(CO$_4$)]$_2$ (**174**; Figure 4.80) [228]; mesityl is 2,4,6-Me$_3$C$_6$H$_2$.

Figure 4.80   The chemical diagram and molecular structure of (**174**)

4.8 Organophosphanes as inverse coordination centres

Organophosphanes offer a great diversity of coordinating options and denticity, making them very important ligands in coordination chemistry. In terms of functioning as inverse coordination centres, these can be constructed about di- and poly-phosphanes. A simple example is found in (μ_2-1,4-phenylenebis(diphenylphosphino)) [Au(benzenethiolato)]$_2$ (**175**; Figure 4.81) [229].

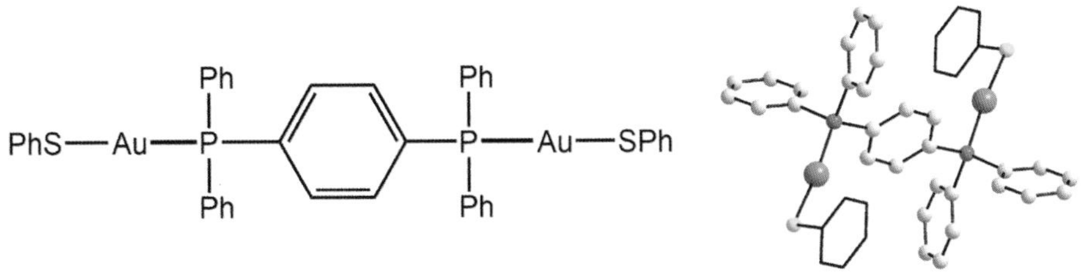

Figure 4.81   The chemical diagram and molecular structure of (**175**)

Linear coordination geometries for gold are also found in the case when the link between the phosphorus donors is biphenyl, as in (μ_2-biphenyl-2,2'-diylbis(diphenylphosphino))[AuCl]$_2$ (**176**; Figure 4.82) [230], and anthracene, as in (μ_2-9,10-bis(diphenylphosphino) anthracene)[AuCl]$_2$ (**177**; Figure 4.83) [231].

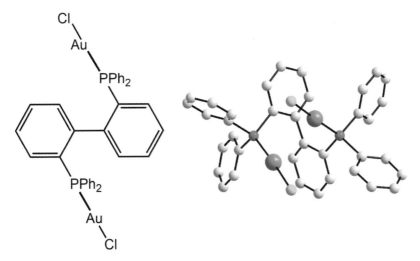

Figure 4.82   The chemical diagram and molecular structure of (**176**)

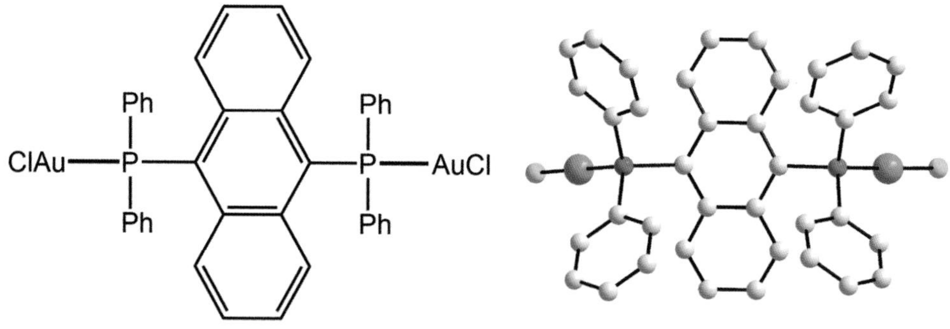

Figure 4.83   The chemical diagram and molecular structure of (**177**)

Diacetylene can also bridge two phosphorus donors as found in the binuclear iron(0) compound (μ_2-1,4-bis(diphenylphosphino) butadiyne)[Fe(CO)$_4$]$_2$ (**178**; Figure 4.84) [232], with a trigonal-bipyramidal geometry about the iron atom.

The link between the phosphorus donors may be an alkyl chain, leading to a non-linear arrangement. This is clearly illustrated by the structure of (μ_2-1,4-bis(diphenylphosphino)butane)[AuCl]$_2$ (**179**; Figure 4.85) [233].

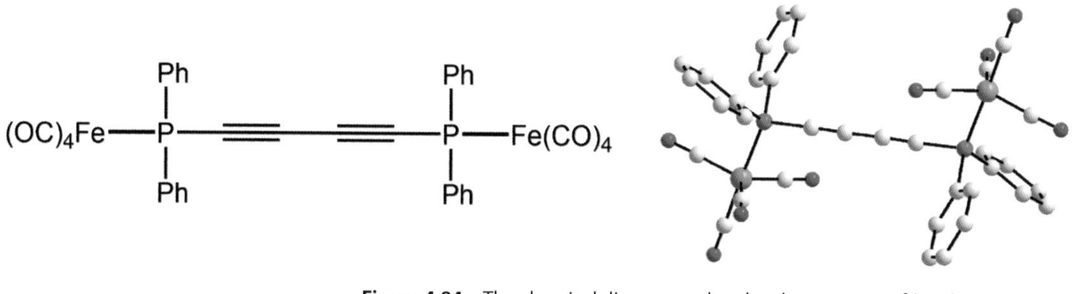

Figure 4.84   The chemical diagram and molecular structure of (**178**)

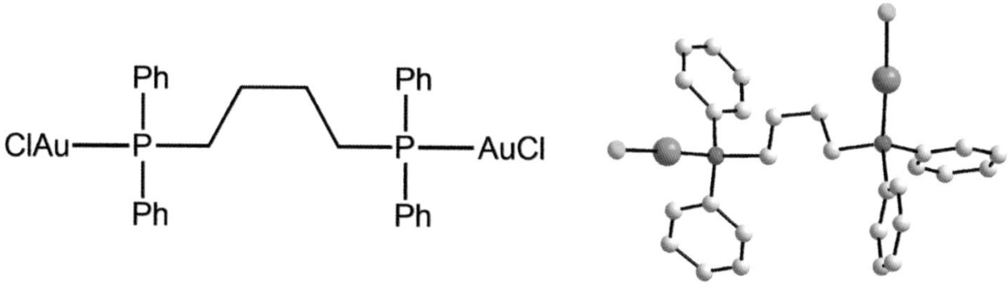

Figure 4.85   The chemical diagram and molecular structure of (**179**)

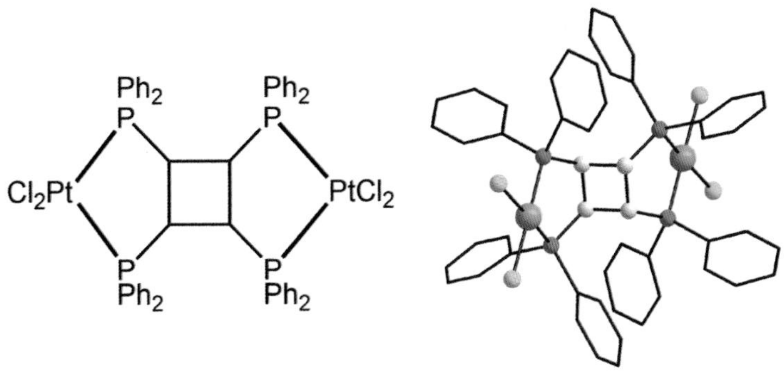

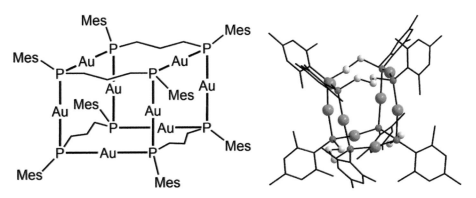

Figure 4.86   The chemical diagram and molecular structure of (**180**)

An example of an inverse coordination centre based on more than two phosphorus connectors is found in the binuclear platinum(II) complex, (tetrakis(diphenylphosphino)cyclobutane) [PtCl$_2$]$_2$ (**180**; Figure 4.86) [224]. Here, the square-planar platinum atoms are chelated by two phosphorus donors.

Another example is found in an attractive octanuclear gold(I) cluster, formulated as (tetrakis(μ_4-propylene-1,3-bis(mesitylphosphino))[Au$_8$] (**181**; Figure 4.87) [235]. Here, the phosphorus atoms carry a formal negative charge so that the molecule is electrically neutral. Each gold atom is linearly coordinated by two phosphorus atoms and each phosphorus atom links two gold atoms.

Figure 4.87   The chemical diagram and molecular structure of (**181**)

4.9 Thiolates and sulphides as inverse coordination centres

Interesting inverse coordination complexes can be derived from aromatic thiols and organic sulphides, as will be demonstrated here. Reflecting the thiophilic behaviour of gold, many of the examples are drawn from gold chemistry. Thus, binuclear gold(I) phosphane derivatives are observed in (μ_2-benzene-1,4-dithiolato) [AuPPh$_2$(2-pyridyl)]$_2$ (182; Figure 4.88) [236] and (μ_2-benzene1,3-dithiolato)[AuPPh$_2$(Me)]$_2$ (183; Figure 4.89) [237].

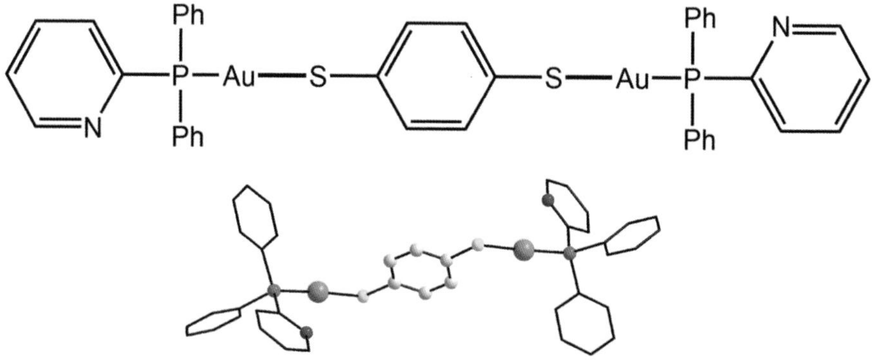

Figure 4.88   The chemical diagram and molecular structure of (**182**)

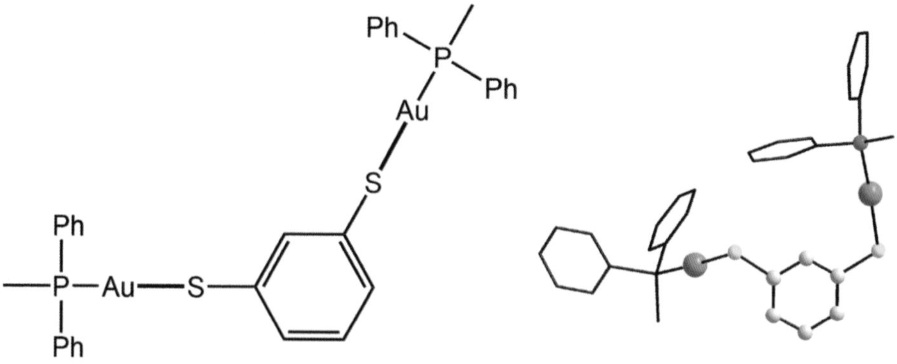

Figure 4.89   The chemical diagram and molecular structure of (**183**)

Reflecting the thiophilic behaviour of gold (as aforementioned) and the flexibility of thiolate-based inversion coordination centres, additional gold moieties can be appended to dithiolates, as shown in (μ_3-benzene-1,3-dithiolato)[AuPPh$_3$]$_3$ (**184**; Figure 4.90) [238], where one of the thiolate-sulphur atoms carries two AuPPh$_3$ substituents.

Three gold(I) phosphane substituents can be appended in trithiolato inversion coordination centres, as exemplified in (μ_3-benzene-1,3,5-trithiolato)[AuPPh$_2$(2-pyridyl)]$_3$ (**185**; Figure 4.91) [236].

With other metals, the related molecules can append substituents via chelating modes as shown in binuclear (μ_2-benzene-1,2,4,5-tetrathiolato)[CoCp*]$_2$ (**186**; Figure 4.92) [239] and in trinuclear (μ_3-benzenehexathiolato)[RhCp*]$_3$ (**187**; Figure 4.93) [240].

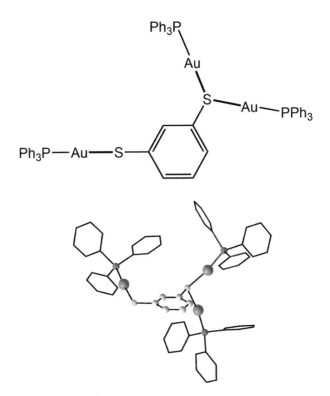

Figure 4.90 The chemical diagram and molecular structure of (**184**)

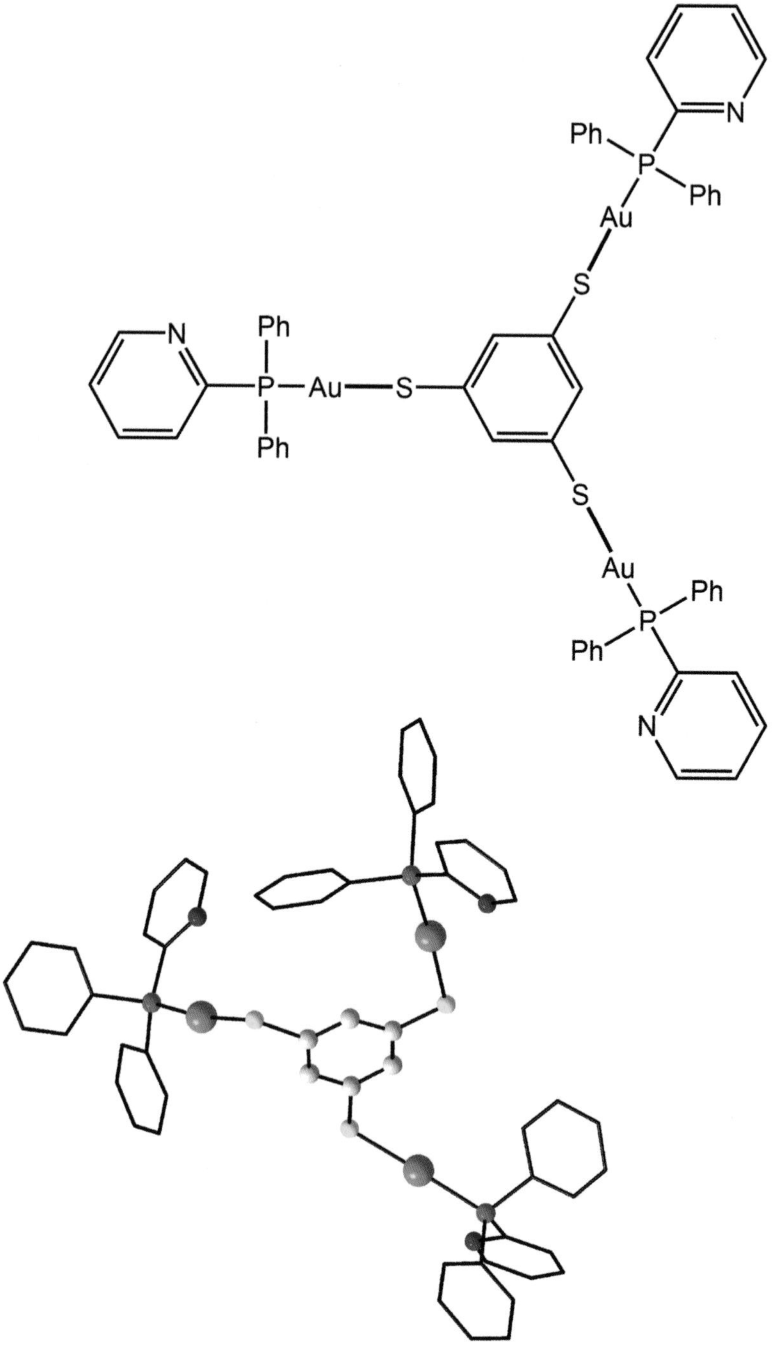

Figure 4.91   The chemical diagram and molecular structure of (**185**)

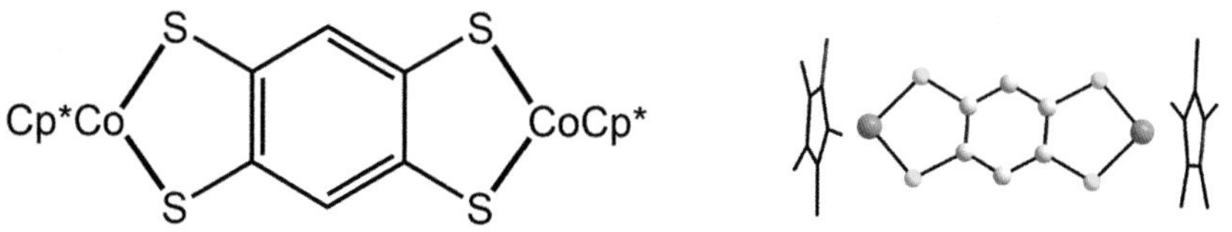

Figure 4.92   The chemical diagram and molecular structure of (**186**)

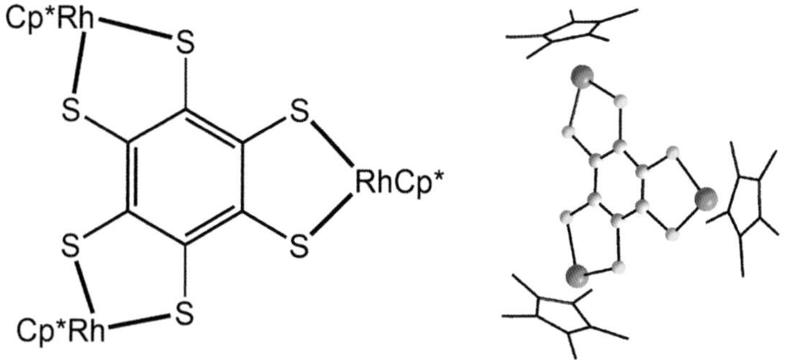

Figure 4.93   The chemical diagram and molecular structure of (**187**)

Not to be outdone, gold(I) forms three-coordinate PS_2 geometries in (μ_6-benzenehexathiolato)[AuPPh$_3$]$_6$ (**188**; Figure 4.94) [241].

Inverse coordination complexes are also formed by thiolates derived from five-membered rings. In (μ_2-1,3-dithiole-2-thione-4,5-dithiolato)[PbPh$_3$]$_2$ (**189**; Figure 4.95) [242], both thiolate-sulphur atoms are attached to tetrahedral lead(IV) centres.

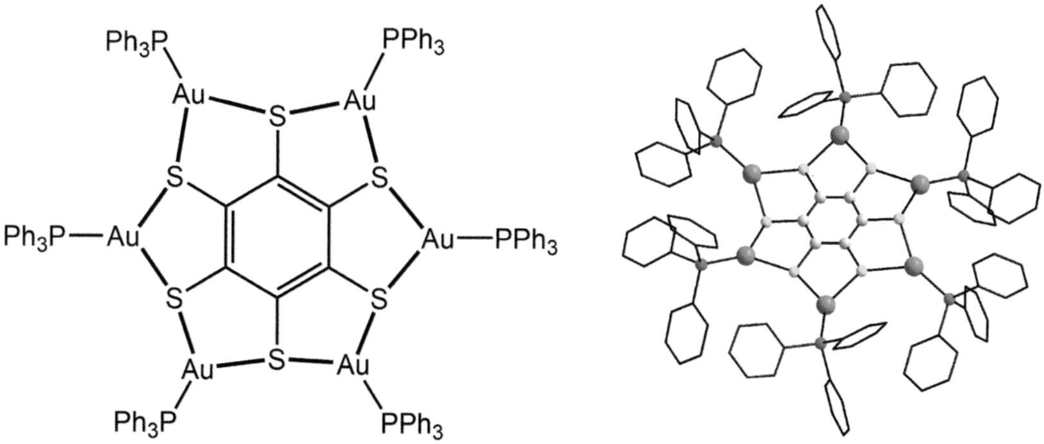

Figure 4.94   The chemical diagram and molecular structure of (**188**)

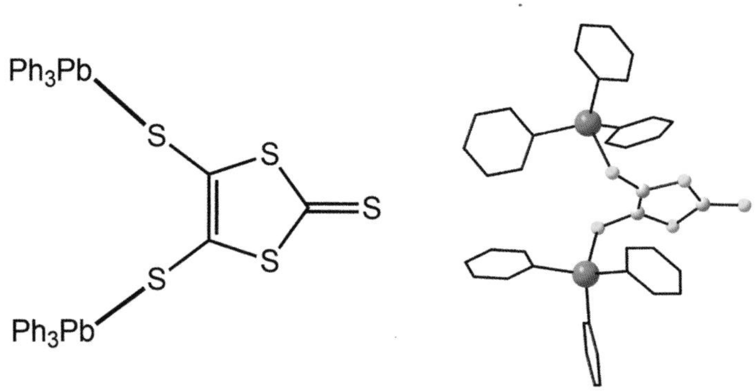

Figure 4.95   The chemical diagram and molecular structure of (**189**)

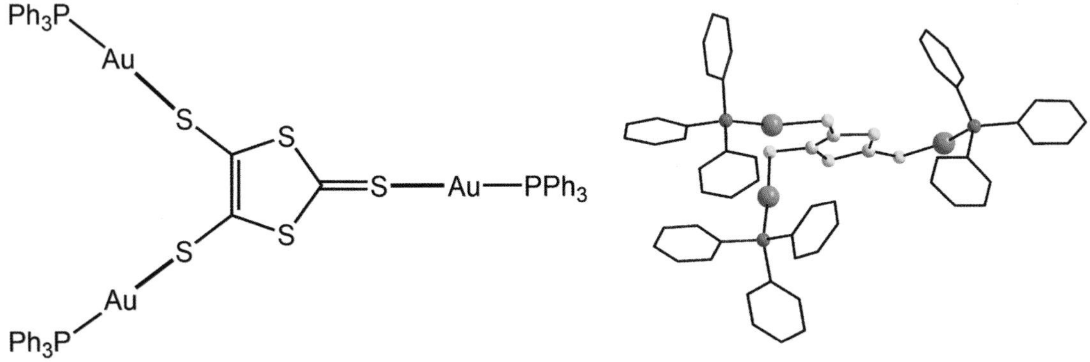

Figure 4.96   The chemical diagram and molecular structure of (**190**)

When gold(I) phosphane is attached, even the thione-sulphur atom engages gold to form a trinuclear monocationic species, $\{(\mu_3\text{-}$ 4,5-dimercapto-1,3-dithiole-2-thione)[AuPPh$_3$]$_3$\}$^+$ (**190**; Figure 4.96) [243], with characteristic linear P–Au–S coordination geometries.

As previously discussed, open-chain organosulphides can function as inverse coordination centres with the structure of (μ_2-(1,2-bis(phenylthio)ethane)[AuCl]$_2$ (**191**; Figure 4.97) [244] being a simple example.

Related organic sulphides with the functional group in the side-chains of an aromatic ring can also form a variety of inverse coordination complexes, as exemplified by binuclear (μ_2-1,2-bis(methylthiomethyl)benzene)[GaCl$_3$]$_2$ (**192**; Figure 4.98) [245], with tetrahedral gallium(III) atoms.

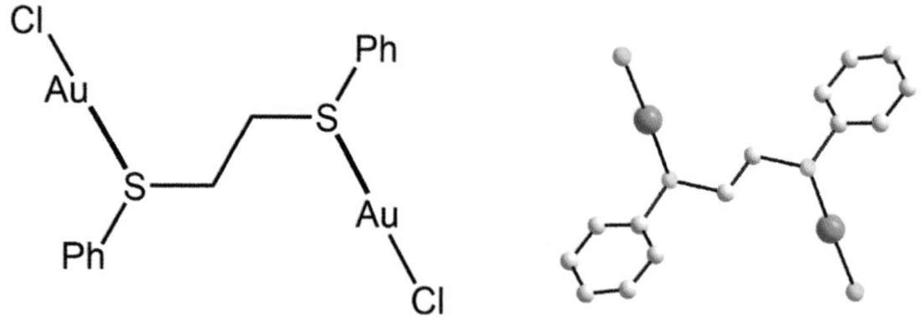

Figure 4.97   The chemical diagram and molecular structure of (**191**)

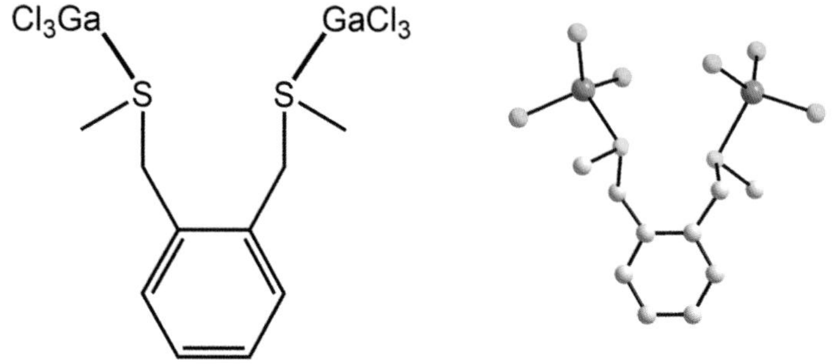

Figure 4.98   The chemical diagram and molecular structure of (**192**)

With four pendent arms, the binuclear inverse coordination complex, dicationic (μ_2-1,2,4,5-tetrakis(t-butylsulfanylmethyl) phenyl))[Pd(NCMe)]$_2$ (**193**; Figure 4.99) [246], can be stabilised with square-planar palladium(II) centres.

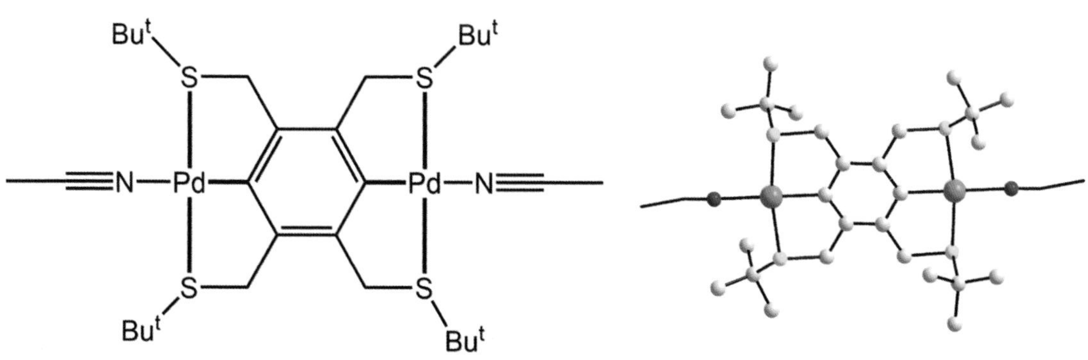

Figure 4.99   The chemical diagram and molecular structure of (**193**)

References

[1]   J. Ribas Gispert, *Coordination Chemistry*, Wiley-VCH, Weinheim, 2008; cf. International Union of Pure and Applied Chemistry (N.G. Connelly, T. Damhus, R.M. Harthorn and A.T. Hutton), *Nomenclature of Inorganic Chemistry: IUPAC Recommendations 2005*, RSC Publishing, Cambridge, 2005.

[2]   R.D. Cannon and R.P. White, *Prog. Inorg. Chem.*, 1988, **36**, 196–297.

[3]   L.E. Orgel, *Nature (London)*, 1960, **187**, 504–505.

[4]   B.N. Figgis and G.B. Robertson, *Nature (London)*, 1965, **205**, 694–695.

[5]   K.J. Schenk and H.U. Güdel, *Inorg. Chem.*, 1982, **21**, 2253–2256.

[6]   I. Haiduc, *Coord. Chem. Rev.*, 2017, **338**, 1–26.

[7]   W. Bragg and G.T. Morgan, *Proc. Roy. Soc. London. Ser.*, 1923, A**104**, 437.

[8]   A. Tulinsky and C.R. Worthington, *Acta Crystallogr.*, 1959, **12**, 626–634.

[9]   J. Ramasamy and J.L. Lambert, *Anal. Chem.*, 1979, **51**, 2044–2045.

[10] R.E. Mulvey, *Chem. Commun.*, 2001, 1049–1056.

[11] W. Clegg, K.W. Henderson, A.R. Kennedy, R.E. Mulvey, C.T. O'Hara, R.B. Rowlings and D.M. Tooke, *Angew. Chem. Int. Ed. Engl.*, 2001, **40**, 3902–3905.

[12] F. García, R.J. Less, M. McPartlin, A. Michalski, R.E. Mulvey, V. Naseri, M.L. Stead, A. Morán de Vega and D.S. Wright, *Chem. Commun.*, 2011, **47**, 1821–1823.

[13] I. Haiduc, *J. Coord. Chem.*, 2018, **71**, 3139–3179.

[14] A.J. Stemmler, J.W. Kampf and V.L. Pecoraro, *Inorg. Chem.*, 1995, **34**, 2271–2272.

[15] G. Mezei, C.M. Zaleski and V.L. Pecoraro, *Chem. Rev.*, 2007, **107**, 4933–5003.

[16] Z. Zheng, C.B. Knobler, M.D. Mortimer, G. Kong and M.F. Hawthorne, *Inorg. Chem.*, 1996, **35**, 1235–1243.

[17] I.H.A. Badr, M. Diaz, M.F. Hawthorne and L.G. Bachas, *Anal. Chem.*, 1999, **71**, 1371–1377.

[18] I. Haiduc and F.T. Edelmann, *Supramolecular Organometallic Chemistry*, Wiley-VCH, Weinheim, 1999.

[19] W.A. Herrmann, *Angew. Chem. Int. Ed. Engl.*, 1986, **25**, 56–76.

[20] M.C. Gimeno and A. Laguna, *Chem. Soc. Rev.*, 2008, **37**, 1952–1966.

[21] I.G. Dance, in *Comprehensive Organometallic Chemistry*, ed. G. Wilkinson, F.G.A. Stone and E.W. Abel, Pergamon, Oxford, 1987, vol. 1, p. 140.

[22] M.N. Sokolov, V.P. Fedin and A.G. Sykes, in *Comprehensive Organometallic Chemistry II*, ed. J.A. McCleverty, T.J. Meyer and A.G. Wedd, Pergamon, Oxford, 2003, vol. 4, p. 762.

[23] I. Haiduc, *Coord. Chem. Rev.*, 2017, **348**, 71–91.

[24] I. Haiduc, *J. Coord. Chem.*, 2019, **72**, 35–52.

[25] C. Oldham, in *Comprehensive Organometallic Chemistry*, ed. G. Wilkinson, F.G.A. Stone and E.W. Abel, Pergamon, Oxford, 1987, vol. 1, p. 441.

[26] C.R. Groom, I.J. Bruno, M.P. Lightfoot and S.C. Ward, *Acta Crystallogr.*, 2016, B**72**, 171–179.

[27] T. Fujihara, J. Aonahata, S. Kumakura, A. Nagasawa, K. Murakami and T. Ito, *Inorg. Chem.*, 1998, **37**, 3779–3784.

[28] K. Anzenhofer and J.J. de Boer, *Rec. Trav. Chim. Pays-Bas*, 1969, **88**, 286.

[29] S.W. Ng, *Acta Crystallogr.*, 2008, E**64**, m1102.

[30] S. Yao, J. Liu and Q. Han, *Acta Crystallogr.*, 2008, E**64**, m989.

[31] W. Bury, E. Chwojnowska, I. Justyniak, J. Lewíski, A. Affek, E. Zygadło-Monikowska, J. Bąk and Z. Florjańczyk, *Inorg. Chem.*, 2012, **51**, 737–745.

[32] K. Mereiter, *Acta Crystallogr.*, 1990, C**46**, 972–976.

[33] M. Ciechanowicz, W.P. Griffith, D. Pawson, A.C. Skapski and M.J. Cleare, *Chem. Commun.*, 1971, 876.

[34] V. Chandrasekhar and R. Thirumoorthi, *Inorg.Chem.*, 2009, **48**, 10330–10337.

[35] H.W. Roesky and I. Haiduc, *J. Chem. Soc., Dalton Trans.*, 1999, 2249–2264.

[36] D.S. Tereshchenko, I.V. Morozov, A.I. Boltalin, E.V. Karpova, T. Yu. Glazunova and S.I. Troyanov, *Crystallogr. Rep.*, 2013, **58**, 68–77.

[37] J. Noack, C. Fritz, C. Flügel, F. Hemmann, H.-J. Gläsel, O. Kahle, C. Dreyer, M. Bauer and E. Kemnitz, *Dalton Trans.*, 2013, **42**, 5706–5710.

[38] J.P.S. Walsh, S.B. Meadows, A. Ghirri, F. Moro, M. Jennings, W.F. Smith, D.M. Graham, T. Kihara, H. Nojiri, I.J. Vitorica-Yrezabal, G.A. Timco, D. Collison, E.J.L. McInnes and R.E.P. Winpenny, *Inorg. Chem.*, 2015, **54**, 12019–12026.

[39] B.F. Straub, F. Rominger and P. Hofmann, *Inorg. Chem.*, 2000, **39**, 2113–2119.

[40] D.M. Ho and R. Bau, *Inorg. Chem.*, 1983, **22**, 4073–4079.

[41] N.W. Alcock, P. Bergamini, T.M. Gomes-Carniero, R.D. Jackson, J. Nicholls, A.G. Orpen, P.G. Pringle, S. Sostero and O. Traverso, *J. Chem. Soc. Chem. Commun.*, 1990, 980–982.

[42] A.E. Underhill and D.M. Watkins, *J. Chem. Soc. Dalton Trans.*, 1977, 5–8.

[43] N. Hoshino, S. Fujita and T. Akutagawa, *Dalton Trans.*, 2019, **48**, 176–181.

[44] W. Uhl and M.R. Halvagar, *Angew. Chem. Int. Ed. Engl.*, 2008, **47**, 1955–1957.

[45] V. Béreau, C.G. Pernin and J.A. Ibers, *Inorg. Chem.*, 2000, **39**, 854–856.

[46] M.A. Halcrow, *Dalton Trans.*, 2009, 2059–2073.

[47] M. Viciano-Chumillas, S. Tanase, L. Jos de Jongh and J. Reedijk, *Eur. J. Inorg. Chem.*, 2010, 3403–3418.

[48] D. Piñero, P. Baran, R. Boca, R. Herchel, M. Klein, R.G. Raptis, F. Renz and Y. Sanakis, *Inorg. Chem.*, 2007, **46**, 10981–10989.

[49] P.A. Angaridis, P. Baran, R. Boča, F. Cervantes-Lee, W. Haase, G. Mezei, R.G. Raptis and R. Werner, *Inorg. Chem.*, 2002, **41**, 2219–2228.

[50] M. Rivera-Camillo, I. Chakraborty, G. Mezei, R.D. Webster and R.G. Raptis, *Inorg. Chem.*, 2008, **47**, 7644–7650.

[51] F. Boutonnet, M. Zablocka, A. Igau, J. Jaud, J.-P. Majoral, J. Schamberger, G. Erker, S. Werner and C. Krüger, *J. Chem. Soc. Chem. Commun.*, 1995, 823–824.

[52] D.T. Richens, L. Helm, P.-A. Pictet, A.E. Merbach, F. Nicolò and G. Chapuis, *Inorg. Chem.*, 1989, **28**, 1394–1402.

[53] S. Ohta, S. Yokozawa, Y. Ohki and K. Tatsumi, *Inorg. Chem.*, 2012, **51**, 2645–2651.

[54] T. Shibahara, G. Sakane and S. Mochida, *J. Am. Chem. Soc.*, 1993, **115**, 10408–10409.

[55] A. Müller and G. Henkel, *Chem. Commun.*, 1996, 1005–1006.

[56] F.A. Cotton, M. Shang and Z.S. Sun, *J. Clust. Sci.*, 1992, **3**, 109–121.

[57] C. Kober, J. Kroner and W. Storch, *Angew. Chem. Int. Ed. Engl.*, 1993, **32**, 1608–1610.

[58] C. von Hanisch and S. Stahl, *Z. Anorg. Allg. Chem.*, 2009, **635**, 2230–2235.

[59] H.W. Roesky, Y. Bai and M. Noltemeyer, *Angew. Chem. Int. Ed. Engl.*, 1989, **28**, 754–755.

[60] J. Lewiński, W. Bury, M. Dutkiewicz, M. Maurin, I. Justyniak and J. Lipkowski, *Angew. Chem. Int. Ed.*, 2008, **47**, 573–576.

[61]   W.T.A. Harrison, M.L.F. Phillips, A.V. Chavez and T.M. Nenoff, *J. Mater. Chem.*, 1999, **9**, 3087–3092.

[62]   S. Menzer, J.R. Phillips, A.M.Z. Slawin, D.J. Williams and J.D. Woollins, *J. Chem. Soc., Dalton Trans.*, 2000, 3269–3273.

[63]   M.A. Malik, P. O'Brien, M. Motevalli and I. Abrahams, *Polyhedron*, 2006, **25**, 241–250.

[64]   A. Albinati, M. Casarin, F. Eisentraeger, C. Maccato, L. Pandolfo and A. Vittadini, *J. Organomet. Chem.*, 2000, **593**, 307–314.

[65]   A. Albinati, M. Casarin, C. Maccato, L. Pandolfo and A. Vittadini, *Inorg. Chem.*, 1999, **38**, 1145–1152.

[66]   S.R.U. Joy, E. Trufan, M.D. Smith, C. Puscas, R.L. Silaghi-Dumitrescu and R.F. Semeniuc, *Inorg. Chim. Acta*, 2019, **485**, 190–199.

[67]   M.K. Ehlert, S.J. Rettig, A. Storr, R.C. Thompson and J. Trotter, *Acta Crystallogr.*, 1994, C**50**, 1023–1026.

[68]   M. Melník, M. Koman and G. Ondrejovič, *Coord. Chem. Rev.*, 2011, **255**, 1581–1586.

[69]   F.A. Cotton, C.A. Murillo and I. Pascual, *Inorg. Chem.*, 1999, **38**, 2746–2749.

[70]   O. Kluge, M. Puidokait, R. Biedermann and H. Krautscheid, *Z. Anorg. Allg. Chem.*, 2007, **633**, 2138–2140.

[71]   G.C. Forbes, A.R. Kennedy, R.E. Mulvey, R.B. Rowlings, W. Clegg, S.T. Liddle and C.C. Wilson, *Chem. Commun.*, 2000, 1759–1760.

[72]   A.R. Kennedy, J.G. MacLellan and R.E. Mulvey, *Acta Crystallogr.*, 2003, C**59**, m302.

[73]   B.F. Abrahams, M.G. Haywood and R. Robson, *Chem. Commun.*, 2004, 938–939.

[74]   M. Driess, J. Aust, K. Merz and C. van Wüllen, *Angew. Chem. Int. Ed.*, 1999, **38**, 3677–3680.

[75]   H.-X. Li, Z.-G. Ren, Y. Zhang, W.-H. Zhang, J.-P. Lang and Q. Shen, *J. Am. Chem. Soc.*, 2005, **127**, 1122–1123.

[76]   E. Barnea, C. Averbuj, M. Kapon, M. Botoshansky and M.S. Eisen, *Eur. J. Inorg. Chem.*, 2007, 4535–4540.

[77]   R.-N. Yang, Y.-A. Sun, Y.-M. Hou, X.-Y. Hu and D.-M. Jin, *Inorg. Chim. Acta*, 2000, **304**, 1–6.

[78]   D. Woodruff, M. Bodensteiner, D.O. Sells, R.E.P. Winpenny and R.A. Layfield, *Dalton Trans.*, 2011, **40**, 10918–10923.

[79]   L.Z. Miller, M. Shatruk and D.T. McQuade, *Chem. Commun.*, 2014, **50**, 8937–8940.

[80]   G. Helgesson, S. Jagner, O. Poncelet and L.G. Hubert-Pfalzgraf, *Polyhedron*, 1991, **10**, 1559–1564.

[81]   L.G. Hubert-Pfalzgraf, S. Daniele, A. Bennaceur, J.-C. Daran and J. Vaissermann, *Polyhedron*, 1997, **16**, 1223–1234.

[82]   D.M. Reis, G.G. Nunes, E.L. Sá, G.R. Friedermann, A.S. Mangrich, D.J. Evans, P.B. Hitchcock, G.J. Leigh and J.F. Soares, *New J. Chem.*, 2004, **28**, 1168–1176.

[83]   C. Knapp, E. Lork, P.G. Watson and R. Mews, *Inorg. Chem.*, 2002, **41**, 2014–2025.

[84]   A.-V. Mudring, T. Timofte and A. Babai, *Inorg. Chem.*, 2006, **45**, 5162–5166.

[85]   Y.P. Jeannin, *Chem. Rev.*, 1998, **98**, 51–76.

[86]   M.T. Pope in *Comprehensive Coordination Chemistry II: Transition Metal Groups 3–6*, ed. J.A. McCleverty and T.J. Meyer, Elsevier Science, New York, 2004, vol. 4, pp. 635–678.

[87]   X. Xie and T. Hughbanks, *Inorg. Chem.*, 2002, **41**, 1824–1830.

[88]   D. Stalke, F.-Q. Liu and H.W. Roesky, *Polyhedron*, 1996, **15**, 2841–2843.

[89]   C. Lorber and L. Vendier, *Inorg. Chem.*, 2013, **52**, 4756–4758.

[90]   L.G. Kuz'mina, A.A. Bagatur'yants, J.A.K. Howard, K.I. Grandberg, A.V. Karchava, E.S. Shubina, L.N. Saitkulova and E.V. Bakhmutova, *J. Organomet. Chem.*, 1999, **575**, 39–50.

[91] H. Schmidbaur, A. Kolb, E. Zeller, A. Schier and H. Beruda, *Z. Anorg. Allg. Chem.*, 1993, **619**, 1575–1579.

[92] J. Vicente, M.-T. Chicote, P. González-Herrero, C. Grünwald and P.G. Jones, *Organometallics*, 1997, **16**, 3381–3387.

[93] B. Kamenar, B. Kaitner and S. Pocev, *J. Chem. Soc. Dalton Trans.*, 1985, 2457–3458.

[94] C.C. Cummins, C. Huang, T.J. Miller, M.W. Reintinger, J.M. Stauber, I. Tannou, D. Tofan, A. Toubaei, A. Velian and G. Wu, *Inorg. Chem.*, 2014, **53**, 3678–3687.

[95] A. Jockisch and H. Schmidbaur, *Z. Naturforsch. B*, 1999, **54b**, 1529–1531.

[96] J. Pickardt, H. Schumann, C.F. Campana and L.F. Dahl, *J. Organomet. Chem.*, 1981, **216**, 245–254.

[97] W. Beck, W. Sacher and U. Nagel, *Angew. Chem.*, 1986, **25**, 270–272.

[98] D. Naumann and F. Schulz, *Z. Anorg. Allg. Chem.*, 2005, **631**, 715–718.

[99] H.J. Breunig, M.A. Mohammed and K.H. Ebert, *Polyhedron*, 1994, **13**, 2471–2472.

[100] A. Jockisch and H. Schmidbaur, *Z. Naturforsch. B*, 1998, **53b**, 1386–1388.

[101] Yu. L. Slovokhotov and Yu. T. Struchkov, *J. Organomet. Chem.*, 1984, **277**, 143–146.

[102] E. Zeller, H. Beruda and H. Schmidbaur, *Chem. Ber.*, 1993, **126**, 2033–2036.

[103] M. Driess, C. Monsé, K. Merz and C. van Wüllen, *Angew. Chem. Int. Ed.*, 2000, **39**, 3684–3686.

[104] F. Canales, C. Gimeno, A. Laguna and M.D. Villacampa, *Inorg. Chim. Acta*, 1996, **244**, 95–103.

[105] J.W. Wielandt, N.L. Kilah, A.C. Willis and S.B. Wild, *Chem. Commun.*, 2006, 3679–3680.

[106] J. von Seyerl, O. Scheidsteger, H. Berke and G. Huttner, *J. Organomet. Chem.*, 1986, **311**, 85–89.

[107] A. Schier, A. Grohmann, J.M. López-de-Luzuriaga and H. Schmidbaur, *Inorg. Chem.*, 2000, **39**, 547–554.

[108] R.E. Bachman and H. Schmidbaur, *Inorg. Chem.*, 1996, **35**, 1399–1401.

[109] E. Zeller and H. Schmidbaur, *J. Chem. Soc. Chem. Commun.*, 1993, 69–70.

[110] Y. Nishibayashi, *Nitrogen Fixation*, Springer Verlag, Berlin, 2017.

[111] B. Peigné and G. Aullón, *Acta Crystallogr.*, 2015, B**71**, 369–386.

[112] S.P. Semproni, D.J. Knobloch, C. Milsmann and P.J. Chirik, *Angew. Chem. Int. Ed.*, 2013, **52**, 5372–5376.

[113] J.-K.F. Buijink, A. Meetsma and J.H. Teuben, *Organometallics*, 1993, **12**, 2004–2005.

[114] R.R. Schrock, R.M. Kolodziej, A.H. Liu, W.M. Davis and M.G. Vale, *J. Am. Chem. Soc.*, 1990, **112**, 4338–4345.

[115] S.P. Semproni and P.J. Chirik, *Eur. J. Inorg. Chem.*, 2013, 3907–3915.

[116] A.M. Madalan, M. Noltemeyer, M. Neculai, H.W. Roesky, M. Schmidtmann, A. Müller, Y. Journaux and M. Andruh, *Inorg. Chim. Acta*, 2006, **359**, 459–467.

[117] L.Y. Goh, C.K. Chu, R.C.S. Wong and T.W. Hambley, *J. Chem. Soc. Dalton Trans.*, 1989, 1951–1956.

[118] L.Y. Goh, R.C.S. Wong and T.C.W. Mak, *J. Organomet.Chem.*, 1989, **364**, 363–371.

[119] M. Di Vaira and L. Sacconi, *Angew. Chem. Int. Ed.*, 1982, **21**, 330–342.

[120] E. Mädl, G. Balázs, E.V. Peresypkina and M. Scheer, *Angew. Chem. Int. Ed.*, 2016, **55**, 7702–7707.

[121] O.J. Scherer, J. Schwalb and G. Wolmershäuser, *New J. Chem.*, 1989, **13**, 399.

[122] L. Weber, U. Sonnenberg, H.-G. Stammler and B. Neumann, *Z. Anorg. Allg. Chem.*, 1991, **605**, 87–99.

[123] M. Caporali, L. Gonsalvi, A. Rossin and M. Peruzzini, *Chem. Rev.*, 2010, **110**, 4178–4235.

[124] P. Barbaro, M. Di Vaira, M. Peruzzini, S.S. Costantini and P. Stoppioni, *Chem.-Eur. J.*, 2007, **13**, 6682–6690.

[125]   F. Spitzer, M. Sierka, M. Latronico, P. Mastrorilli, A.V. Virovets and M. Scheer, *Angew. Chem. Int. Ed.*, 2015, **54**, 4392–4396.

[126]   C. Schwarzmaier, A.Y. Timoshkin, G. Balázs and M. Scheer, *Angew. Chem. Int. Ed.*, 2014, **53**, 9077–9081.

[127]   M. Scheer, C. Troitzsch, L. Hilfert, M. Dargatz, E. Kleinpeter, P.G. Jones and J. Sieler, *Chem. Ber.*, 1995, **128**, 251–257.

[128]   O.J. Scherer, M. Swarowsky and G. Wolmershäuser, *Organometallics*, 1989, **8**, 841–842.

[129]   M.E. Barr, S.K. Smith, B. Spencer and L.F. Dahl, *Organometallics*, 1991, **10**, 3983–3991.

[130]   F. Kraus, T. Hanauer and N. Korber, *Inorg. Chem.*, 2006, **45**, 1117–1123.

[131]   A.S.P. Frey, F.G.N. Cloke, P.B. Hitchcock and J.C. Green, *New J. Chem.*, 2011, **35**, 2022–2026.

[132]   M. Scheer, M. Dargatz, K. Schenzel and P.G. Jones, *J. Organomet. Chem.*, 1992, **435**, 123–132.

[133]   R. Hoffmann, *Angew. Chem. Int. Ed.*, 1982, **21**, 711–724.

[134]   L.Y. Goh, R.C.S. Wong, C.K. Chu and T.W. Hambley, *J. Chem. Soc. Dalton Trans.*, 1990, 977–982.

[135]   O.J. Scherer, T. Brück and G. Wolmershäuser, *Chem. Ber.*, 1989, **122**, 2049–2054.

[136]   M. Fleischmann, C. Heindl, M. Seidl, G. Balázs, A.V. Virovets, E.V. Peresypkina, M. Tsunoda, F.P. Gabbaï and M. Scheer, *Angew. Chem. Int. Ed.*, 2012, **51**, 9918–9921.

[137]   O.J. Scherer, H. Swarowsky, G. Wolmershäuser, W. Kaim and S. Kohlmann, *Angew. Chem. Int. Ed.*, 1987, **26**, 1153–1155.

[138]   O.J. Scherer, B. Werner, G. Heckmann and G. Wolmershäuser, *Angew. Chem. Int. Ed.*, 1991, **30**, 553–555.

[139]   D. Weber, C. Mujica and H.G. von Schnering, *Angew. Chem. Int. Ed.*, 1982, **21**, 863.

[140]   S. Charles, J.A. Danis, J.C. Fettinger and B.W. Eichhorn, *Inorg. Chem.*, 1997, **36**, 3772–3778.

[141]   S.N. Konchenko, N.A. Pushkarevsky, M.T. Gamer, R. Köppe, H. Schnöckel and P.W. Roesky, *J. Am. Chem. Soc.*, 2009, **131**, 5740–5741.

[142]   H. Krautscheid, E. Matern, J. Olkowska-Oetzel, J. Pikies and G. Fritz, *Z. Anorg. Allg. Chem.*, 2001, **627**, 2118–2120.

[143]   F. Dielmann, M. Sierka, A.V. Virovets and M. Scheer, *Angew. Chem. Int. Ed.*, 2010, **49**, 6860–6864.

[144]   F. Dielmann, A. Timoshkin, M. Piesch, G. Balázs and M. Scheer, *Angew. Chem. Int. Ed.*, 2017, **56**, 1671–1675.

[145]   G. Fritz, H.-W. Schneider, W. Hönle and H.G. von Schnering, *Z. Anorg. Allg. Chem.*, 1990, **585**, 51–64.

[146]   O.J. Scherer, T. Völmecke and G. Wolmershäuser, *Eur. J. Inorg. Chem.*, 1999, 945–949.

[147]   T. Li, M.T. Gamer, M. Scheer, S.N. Konchenko and P.W. Roesky, *Chem. Commun.*, 2013, **49**, 2183–2185.

[148]   B. Modec, M. Šala and R. Clérac, *Eur. J. Inorg. Chem.*, 2010, 542–553.

[149]   J.R. Kirchhoff and K. Kirschbaum, *Polyhedron*, 1998, **17**, 4033–4039.

[150]   A. Neels, B.M. Neels, H. Stoeckli-Evans, A. Clearfield and D.M. Poojary, *Inorg. Chem.*, 1997, **36**, 3402–3409.

[151]   K. Sakai and M. Kurashima, *Acta Crystallogr.*, 2003, **E59**, m411–m413.

[152]   M. Ferigo, P. Bonhôte, W. Marty and H. Stoeckli-Evans, *J. Chem. Soc. Dalton Trans.*, 1994, 1549–1554.

[153]   A.H. Pedersen, M. Julve, J. Martínez-Lillo, J. Cano and E.K. Brechin, *Dalton Trans.*, 2017, **46**, 11890–11897.

[154]   N. Nédélec and F.D. Rochon, *Inorg. Chem.*, 2001, **40**, 5236–5244.

[155]   P. von Grebe, P.J.S. Miguel and B. Lippert, *Z. Anorg. Allg. Chem.*, 2012, **638**, 1691–1698.

[156]   Y.E. Türkmen, S. Sen and V.H. Rawal, *CrystEngComm*, 2013, **15**, 4221–4224.

[157]   W. Kaufman, L.M. Venanzi and A. Albinati, *Inorg. Chem.*, 1988, **27**, 1178–1187.

[158]   M. Trivedi, D.S. Pandey, R.-Q. Zou and Q. Xu, *Inorg. Chem. Commun.*, 2008, **11**, 526–530.

[159]   S. Tampucci, M.B. Ferrari, L. Calucci, G. Pelosi and G. Denti, *Inorg. Chim. Acta*, 2007, **360**, 2814–2818.

[160]   K. Ha, *Z. Kristallogr. – New Cryst. Struct.*, 2011, **226**, 313–314.

[161]   C.S. Lai and E.R.T. Tiekink, *CrystEngComm*, 2004, **6**, 593–605.

[162]   R. Haiges, P. Deokar and K.O. Christe, *Angew. Chem. Int. Ed.*, 2014, **53**, 5431–5434.

[163]   N. Soltanzadeh and A. Morsali, *Polyhedron*, 2009, **28**, 1343–1347.

[164]   D. Ogrin, L.H. van Poppel, S.G. Bott and A.R. Barron, *Dalton Trans.*, 2004, 3689–3694.

[165]   H.D. Arman, P. Poplaukhin and E.R.T. Tiekink, *Acta Crystallogr.*, 2009, **E65**, m1472–m1473.

[166]   M. Jordan, W. Saak, D. Haase and R. Beckhaus, *Organometallics*, 2010, **29**, 5859–5870.

[167]   T.D. Hamilton, G.S. Papaefstathiou and L.R. MacGillivray, *CrystEngComm*, 2002, **4**, 223–226.

[168]   A. Okazawa, T. Ishida and T. Nogami, *Chem. Lett.*, 2004, **33**, 1478–1479.

[169]   A.W. Kleij, M. Lutz, A.L. Spek, P.W.N.M. van Leeuwen and J.N.H. Reek, *Chem. Commun.*, 2005, 3661–3663.

[170]   F.A. Cotton, E.V. Dikarev, M.A. Petrukhina, M. Schmitz and P.J. Stang, *Inorg. Chem.*, 2002, **41**, 2903–2908.

[171]   V. Shuster, S. Gambarotta, G.B. Nikiforov, I. Korobkov and P.H.M. Budzelaar, *Organometallics*, 2012, **31**, 7011–7019.

[172]   B.Z. Momeni, N. Fathi, M. Shafiei, F. Ghasemi and F. Rominger, *J. Coord. Chem.*, 2016, **69**, 2697–2706.

[173]   S. Kraft, R. Beckhaus, D. Haase and W. Saak, *Angew. Chem. Int. Ed.*, 2004, **43**, 1583–1587.

[174]   Q. Zhao, R.-F. Li, S.-K. Xing, X.-M. Liu, T.-L. Hu and X.-H. Bu, *Inorg. Chem.*, 2011, **50**, 10041–10046.

[175]   T. Brietzke, D. Kässler, A. Kelling, U. Schilde and H.-J. Holdt, *Acta Crystallogr.*, 2014, **E70**, m39–m40.

[176]   S.D. Bergman, I. Goldberg, A. Barbieri, F. Barigelletti and M. Kol, *Inorg. Chem.*, 2004, **43**, 2355–2367.

[177]   X. Ma, Y. Pan, L. Wu, X. Huang, H. Sun and C. Zhu, *Acta Crystallogr.*, 1997, **C53**, 278–279.

[178]   A. Venugopal, I. Kamps, D. Bojer, R.J.F. Berger, A. Mix, A. Willner, B. Neumann, H.-G. Stammler and N.W. Mitzel, *Dalton Trans.*, 2009, 5755–5765.

[179]   E. Yang, R.-Q. Zhuang and Y.-E. Chen, *Acta Crystallogr.*, 2006, **E62**, m2901–m2903.

[180]   C.A. Ellis and E.R.T. Tiekink, *Acta Crystallogr.*, 2006, **E62**, m3049–m3051.

[181]   A.M. Kirillov, *Coord. Chem. Rev.*, 2011, **255**, 1603–1622.

[182]   J.B. Hill, S.J. Eng, W.T. Pennington and G.H. Robinson, *J. Organomet. Chem.*, 1993, **445**, 11–18.

[183]   L. Bolundut, I. Haiduc, M.F. Mahon and K.C. Molloy, *Rev. Chim. (Bucureşti)*, 2008, **59**, 1194–1196.

[184]   C. Chopard, C. Lenoir, S. Rizzato, M. Vidal, J. Arpalahti, L. Gabison, A. Albinati, C. Garbay and J. Kozelka, *Inorg. Chem.*, 2008, **47**, 9701–9705.

[185]   O. Crespo, M.C. Gimeno, P.G. Jones, A. Laguna, M. Naranjo and M.D. Villacampa, *Eur. J. Inorg. Chem.*, 2008, 5408–5417.

[186]   R.N. Patel, N. Singh, K.K. Shukla, V.L.N. Gundla and U.K. Chauhan, *J. Inorg. Biochem.*, 2005, **99**, 651–663.

[187]   J.W. Slater, D.M. D'Alessandro, F.R. Keene and P.J. Steel, *Dalton Trans.*, 2006, 1954–1962.

[188]   R.-S. Zhou and J.-F. Song, *Acta Crystallogr.*, 2009, **E65**, m1523.

[189]   M.S. Haddad, E.N. Duesler and D.N. Hendrickson, *Inorg. Chem.*, 1979, **18**, 141–148.

[190] P. Kircher, G. Huttner, K. Heinze, B. Schiemenz, L. Zsolnai, M. Büchner and A. Driess, *Eur. J. Inorg. Chem.*, 1998, 703–720.

[191] P.H. Dinolfo, K.D. Benkstein, C.L. Stern and J.T. Hupp, *Inorg. Chem.*, 2005, **44**, 8707–8714.

[192] B.-C. Tzeng, D. Li, S.-M. Peng and C.-M. Che, *J. Chem. Soc. Dalton Trans.*, 1993, 2365–2371.

[193] M. Serratrice, M.A. Cinellu, L. Maiore, M. Pilo, A. Zucca, C. Gabbiani, A. Guerri, I. Landini, S. Nobili, E. Mini and L. Messori, *Inorg. Chem.*, 2012, **51**, 3161–3171.

[194] S. Zhang, W. Shi and P. Cheng, *Coord. Chem. Rev.*, 2017, **352**, 108–150.

[195] G. Net, J.C. Bayon, P. Esteban, P.G. Rasmussen, A. Alvarez-Larena and J.F. Piniella, *Inorg. Chem.*, 1993, **32**, 5313–5321.

[196] W.-J. Tai, C.-Y. Li, P.-H. Lin, J.-Y. Li, M.-J. Chen and B.-T. Ko, *Appl. Organomet. Chem.*, 2012, **26**, 518–527.

[197] J.-L. Chen, X.-H. Zeng, Y.-S. Luo, W.-M. Wang, L.-H. He, S.-J. Liu, H.-R. Wen, S. Huang, L. Liu and W.-Y. Wong, *Dalton Trans.*, 2017, **46**, 13077–13087.

[198] A. Tăbăcaru, C. Pettinari and S. Galli, *Coord. Chem. Rev.*, 2018, **372**, 1–30.

[199] A. Takayama, T. Suzuki, M. Ikeda, Y. Sunatsuki and M. Kojima, *Dalton Trans.*, 2013, **42**, 14556–14567.

[200] A.J. Mota, A. Rodríguez-Diéguez, M.A. Palacios, J.M. Herrera, D. Luneau and E. Colacio, *Inorg. Chem.*, 2010, **49**, 8986–8996.

[201] A. Rodríguez-Diéguez, A.J. Mota, J.M. Seco, M.A. Palacios, A. Romerosa and E. Colacio, *Dalton Trans.*, 2009, 9578–9586.

[202] H. Langhals, M. Limmert, I.-P. Lorenz, P. Mayer, H. Piotrowski and K. Polborn, *Eur. J. Inorg. Chem.*, 2000, 2345–2349.

[203] S. Zhang, W. Shi and P. Cheng, *Coord. Chem. Rev.*, 2017, **352**, 108–150.

[204] G. Smith, A.N. Reddy, K.A. Byriel and C.H.L. Kennard, *J. Chem. Soc., Dalton Trans.*, 1995, 3565–3570.

[205] J. Cano, G. De Munno, J.L. Sanz, R. Ruiz, J. Faus, F. Lloret, M. Julve and A. Caneschi, *J. Chem. Soc., Dalton Trans.*, 1997, 1915–1924.

[206] X. Xiao, D. Du, M. Tian, X. Han, J. Liang, D. Zhu and L. Xu, *J. Organomet. Chem.*, 2012, **715**, 54–63.

[207] B. Tao, H. Xia, C.-X. Huang and X.-W. Li, *Z. Anorg. Allg. Chem.*, 2011, **637**, 703–707.

[208] D. Dakternieks, D.J. Clarke and E.R.T. Tiekink, *Z. Kristallogr. – Cryst. Mater.*, 2002, **217**, 622–626.

[209] X. Xiao, D. Du, X. Han, J. Liang, M. Tian, D. Zhu and L. Xu, *J. Organomet. Chem.*, 2012, **713**, 143–150.

[210] C. Zhuang, N. Li and X.-Y. Yu, *Acta Crystallogr.*, 2012, E**68**, m268–m269.

[211] R. West, *Oxocarbons*, Academic Press, New York, 1980.

[212] O.T. Summerscales, F.G.N. Cloke, P.B. Hitchcock, J.C. Green and N. Hazari, *Science*, 2006, **311**, 829–831.

[213] R. Lalrempuia, C.E. Kefalidis, S.J. Bonyhady, B. Schwarze, L. Maron, A. Stasch and C. Jones, *J. Am. Chem. Soc.*, 2015, **137**, 8944–8947.

[214] I. Castro, M.L. Calatayud, J. Sletten, F. Lloret and M. Julve, *Inorg. Chim. Acta*, 1999, **287**, 173–180.

[215] G. Bernardinelli, D. Deguenon, R. Soules and P. Castan, *Can. J. Chem.*, 1989, **67**, 1158–1165.

[216] I. Castro, J. Sletten, J. Faus, M. Julve, Y. Journaux, F. Lloret and S. Alvarez, *Inorg. Chem.*, 1992, **31**, 1889–1894.

[217] T.-W. Tseng, T.-T. Luo, S.-H. Liao, K.-H. Lu and K.-L. Lu, *Angew. Chem., Int. Ed.*, 2016, **55**, 8343–8347.

[218] J.V. Folgado, R. Ibáñez, E. Coronado, D. Beltrán, J.M. Savariault and J. Galy, *Inorg. Chem.*, 1988, **27**, 19–26.

[219] P. Thuéry, *Cryst. Growth Des.*, 2014, **14**, 2665–2676.

[220] F. Gerstner, W. Schwarz, H.-D. Hausen and J. Weidlein, *J. Organomet. Chem.*, 1979, **175**, 33–47.

[221] D. Walther, M. Stollenz, L. Böttcher and H. Görls, *Z. Anorg. Allg. Chem.*, 2001, **627**, 1560–1570.

[222] G.A. Holloway and T.B. Rauchfuss, *Inorg. Chem.*, 1999, **38**, 3018–3019.

[223] M. Hayashi, K. Otsubo, T. Kato, K. Sugimoto, A. Fujiwara and H. Kitagawa, *Chem. Commun.*, 2015, **51**, 15796–15799.

[224] B. Ramakrishna, R. Nagarajaprakash and B. Manimaran, *J. Organomet. Chem.*, 2015, **791**, 322–327.

[225] F. Lloret, J. Sletten, R. Ruiz, M. Julve, J. Faus and M. Verdaguer, *Inorg. Chem.*, 1992, **31**, 778–784.

[226] B. Wenzel, B. Wehse, U. Schilde and P. Strauch, *Z. Anorg. Allg. Chem.*, 2004, **630**, 1469–1476.

[227] S. Lanza, F. Nicolò, G. Cafeo, H.A. Rudbari and G. Bruno, *Inorg. Chem.*, 2010, **49**, 9236–9246.

[228] D. Walther, S. Liesicke, L. Böttcher, R. Fischer, H. Görls and G. Vaughan, *Inorg. Chem.*, 2003, **42**, 625–632.

[229] I.O. Koshevoy, E.S. Smirnova, M. Haukka, A. Laguna, J.C. Chueca, T.A. Pakkanen, S.P. Tunik, I. Ospino and O. Crespo, *Dalton Trans.*, 2011, **40**, 7412–7422.

[230] M.J. Mayoral, P. Ovejero, R. Criado, M.C. Lagunas, A. Pintado-Alba, M.R. Torres and M. Cano, *J. Organomet. Chem.*, 2011, **696**, 2789–2796.

[231] K. Zhang, J. Prabhavathy, J.H.K. Yip, L.L. Koh, G.K. Tan and J.J. Vittal, *J. Am. Chem. Soc.*, 2003, **125**, 8452–8453.

[232] C.J. Adams, M.I. Bruce, E. Horn and E.R.T. Tiekink, *J. Chem. Soc., Dalton Trans.*, 1992, 1157–1164.

[233] H. Schmidbaur, P. Bissinger, J. Lachmann and O. Steigelmann, *Z. Naturforsch. B*, 1992, **47b**, 1711–1716.

[234] T. Stampfl, R. Gutmann, G. Czermak, C. Langes, A. Dumfort, H. Lopacka, K.-H. Ongania and P. Brüggeller, *Dalton Trans.*, 2003, 3425–3435.

[235] E.M. Lane, T.W. Chapp, R.P. Hughes, D.S. Glueck, B.C. Feland, G.M. Bernard, R.E. Wasylishen and A.L. Rheingold, *Inorg. Chem.*, 2010, **49**, 3950–3957.

[236] K. Nunokawa, K. Okazaki, S. Onaka, M. Ito, Y. Sunahara, T. Ozeki, H. Imai and K. Inoue, *J. Organomet. Chem.*, 2005, **690**, 1332–1339.

[237] M.C. Gimeno, P.G. Jones, A. Laguna, M. Laguna and R. Terroba, *Inorg. Chem.*, 1994, **33**, 3932–3938.

[238] H. Ehlich, A. Schier and H. Schmidbaur, *Inorg. Chem.*, 2002, **41**, 3721–3727.

[239] M. Nomura and M. Fourmigué, *Inorg. Chem.*, 2008, **47**, 1301–1312.

[240] Y. Shibata, B. Zhu, S. Kume and H. Nishihara, *Dalton Trans.*, 2009, 1939–1943.

[241] H.K. Yip, A. Schier, J. Riede and H. Schmidbaur, *J. Chem. Soc., Dalton Trans.*, 1994, 2333–2334.

[242] S.M.S.V. Doidge-Harrison, J.T.S. Irvine, G.M. Spencer, J.L. Wardell, P. Ganis, G. Valle and G. Tagliavini, *Polyhedron*, 1996, **15**, 1807–1815.

[243] E. Cerrada, A. Laguna, M. Laguna and P.G. Jones, *J. Chem. Soc., Dalton Trans.*, 1994, 1325–1326.

[244] M.G.B. Drew and M.J. Riedl, *J. Chem. Soc., Dalton Trans.*, 1973, 52–55.

[245] C. Gurnani, W. Levason, R. Ratnani, G. Reid and M. Webster, *Dalton Trans.*, 2008, 6274–6282.

[246] **S.J. Loeb, G.K.H. Shimizu and J.A. Wisner, *Organometallics*, 1998, 17, 2324–2327.**

Index